W9-CID-035

Muscles

Muscles:

The Magic of Motion

By Robert D. Selim
and the Editors of U.S.News Books

U.S.NEWS BOOKS Washington, D.C.

U.S.NEWS BOOKS

THE HUMAN BODY
Muscles: The Magic of Motion

Series Consultants

Donald M. Engelman is Molecular Biophysicist and Biochemist at Yale University and a guest Biophysicist at the Brookhaven National Laboratory in New York. A specialist in biological structure, Dr. Engelman has published research in American and European journals. From 1976 to 1980, he was chairman of the Molecular Biology Study Section at the National Institutes of Health.

Stanley Joel Reiser is Associate Professor of Medical History at Harvard Medical School and codirector of the Kennedy Interfaculty Program in Medical Ethics at the University. He is the author of *Medicine and the Reign of Technology* and coeditor of *Ethics in Medicine: Historical Perspectives and Contemporary Concerns.*

Harold C. Slavkin, Professor of Biochemistry at the University of Southern California, directs the Graduate Program in Craniofacial Biology and also serves as Chief of the Laboratory for Developmental Biology in the University's Gerontology Center. His research on the genetic basis of congenital defects of the head and neck has been widely published.

Lewis Thomas is Chancellor of the Memorial Sloan-Kettering Cancer Center in New York City. A member of the National Academy of Sciences, Dr. Thomas has served on advisory councils of the National Institutes of Health. He has written *The Medusa and the Snail* and *The Lives of a Cell,* which received the 1974 National Book Award in Arts and Letters.

Special Consultant for Muscles

Gideon Ariel, a former Olympic athlete, is widely known for applying scientific principles to the improvement of athletic performance. He is president of the Coto Research Center in Trabuco Canyon, California, an organization specializing in the measurement of motion and mechanical performance.

Consultants for Muscles

John V. Basmajian is Professor of Medicine and Director of Rehabilitation Programs at the McMaster University School of Medicine, Chedoke-McMaster Hospitals, in Hamilton, Ontario, Canada. A clinical neurophysiologist, he is also a specialist in electromyography, the study of contraction and electrical properties of muscles, and kinesiology, the study of motion. Dr. Basmajian is the author of twelve books on muscle and more than 300 scholarly articles. In 1977, he was awarded the Gold Key by the American Congress of Rehabilitation Medicine, the highest honor in the field of muscle rehabilitation.

Edward V. Evarts is Chief of the Laboratory of Neurophysiology of the National Institute of Mental Health in Bethesda, Maryland. His principal fields of research are the neurophysiology of movement and the effects of neurological disease on muscles and motor function. He has studied the role of the brain in movement and has developed techniques of measuring the activity of brain neurons during movement. Dr. Evarts serves on the editorial boards of several major technical journals and has received numerous national and international scientific awards.

Ernst Jokl is Clinical Professor of Neurology at the University of Kentucky, Lexington, Kentucky, and Honorary Professor of Sports Medicine at the Universities of Berlin and Frankfort, Germany. A founder of sports medicine in the United States, he has written many books on the medical aspects of sports. He has served as Fulbright Professor of Medicine of the U.S. State Department and as adviser to Vice President Hubert Humphrey. Dr. Jokl is also Honorary President of the Research Committee of the International Council of Sport and Physical Education of the United Nations.

Picture Consultants

Amram Cohen is General Surgery Resident at the Walter Reed Army Medical Center in Washington, D.C.

Richard G. Kessel, Professor of Zoology at the University of Iowa, studies cells, tissues and organs with scanning and transmission electron microscopy instruments. He is coauthor of two books on electron microscopy.

U.S.News Books, a division of U.S.News & World Report, Inc.

Library of Congress Cataloging in Publication Data

Selim, Robert D.
Muscles: the magic of motion.
(The Human body)
Includes index.
1. Human mechanics. 2. Muscles. I. U.S. News Books. II. Title. III. Series.
QP303.S44 1982 612'.7 82-7050
AACR2
ISBN 0-89193-607-6
ISBN 0-89193-637-8 (leatherbound)
ISBN 0-89193-667-X (school ed.)

20 19 18 17 16 15 14 13 12 11
10 9 8 7 6 5 4 3 2 1

Contents

Introduction:

It Seems to Dance

Tautly drawn and tightly packed, fibers of skeletal muscle, magnified 140 times, teem with tense energy. Muscle fibers, as pure in action as in design, move only in the straight lines of their construction. Yet by simply contracting, muscle fibers produce a vast array of motions to enliven the human form with speed, grace and power.

Pirouetting about the sun — day by day, season by season — the earth keeps time for everything upon it. And everything, quickened by diverse forces and tripping to distinct tunes, expresses itself in movement. With a mind as lively as his body, man is moved as much by marvel as muscle. Sir John Davies, a contemporary of Shakespeare, rhymed:

> For your quick eyes in wandering to and fro,
> From east to west, on no one thing can glance,
> But, if you mark it well, it seems to dance.

Man fathoms the nature of things by tracking their motion. All motion follows mechanical principles. Like machines he makes, man is a set of levers whose movements copy the geometry of classical mechanics. These levers are powered by muscles, elegant, efficient machines, their action as simple as their character is complex. Against the motions they create, muscles seem awkward. They merely contract. But this prosaic operation belies a division of labor so elaborate scientists have just begun to unravel it.

Each of more than six hundred muscles is shot through with nerves. Linking muscles to the brain and spinal cord, labyrinthian circuits hum with signals directing the ebb and flow of muscular energy. Many muscles work together to perform the simplest tasks. And much muscular activity occurs beyond the ken of the conscious mind as the body, through its neuromuscular network, manages its own motion.

Like any source of power, the work of muscles may be impaired or enhanced. By mastering muscles' mysteries, scientists raise promise of progress in treating diseases that hamper movement. Likewise, as athletes train to more refined regimens they reach ever-higher standards of achievement. Power and grace eloquently express the vitality and beauty of the magnificent, mindful machine called man.

Chapter 1

The Body in Motion

Motion signals life as surely as an untimely twitch betrays a cornered opossum to a foraging fox. In a universe where everything moves, everything seems alive. Heavenly bodies spin as they wheel about the skies. Seas swell and rivers rush. Branches sway, blossoms open and leaves fall. Animals walk and fly, swim, slither and burrow around the world. Arrows seem to spring from bowstrings, and shot to burst from gun barrels. Medieval scholars and Renaissance poets pictured the universe as a dance, a cosmic light fantastic tripped to the strains of the heavenly spheres.

A lead dancer in this myriad *corps de ballet,* man has a fluid body and a restless mind. Movement and contemplation, seemingly so different, are but two sides of one coin. Motion fulfills ambition and expresses emotion and ideals. Minds are supple — jumping to conclusions, groping for answers, wrestling with dilemmas. Like Mercury, the swift, cunning messenger of Zeus, man pulses with quicksilver. As seventeenth-century English poet Henry Vaughan rhymed:

> He hath no root, nor to one place is ty'd
> But ever restless and Irregular
> About this Earth doth run and ride.

Few motions of the mind are more relentless than the curiosity which drives man to understand the world and his place in it.

Comings and goings of things around him have forever whetted his curiosity. Science grew from efforts to explain how and why things move. The universe was long conceived of as a living organism, created by divine purpose and animated by divine will. Things, inanimate and animate alike, moved because God willed that they do so, much as a pianist wills his fingers over the keyboard. With the scientific revolution of the sixteenth and seventeenth centuries, the universe came to be conceived of as a complex machine. Designed,

Flashing like the spokes of a spinning wheel, American gymnastic champion Kurt Thomas whips his body around the pommel horse with mechanical symmetry and angelic grace. A set of bony, sinewy hinges and levers driven by muscles, man moves like a machine but one with willful purpose.

Startled by danger, impalas bolt for cover in a seemingly aimless frenzy of leaps and bounds. Jumping as far as twenty-five feet, they easily clear obstacles and bewilder pursuers with erratic flight.

Fleeing without flying, the flightless ostrich carries its 300 pounds at thirty miles per hour, outrunning every other animal on two legs. When cornered, its powerful legs can deal damaging blows.

wound and serviced by God, this machine was nevertheless driven by forces measurable to man. Mysteries of celestial motion fell first to the new science of mechanics. Its theorems and formulas followed the trail of living things. Paradoxically, only when animals were treated as machines instead of organisms did the magic of their motion begin to yield its secrets.

Like machines, animals can only move forward by pushing themselves backward against their surroundings. Just as cars, ships and planes use wheels, screws and propellers, so animals use legs, fins and wings to press backward against land, water and air. In every case, the force available for moving something forward equals the force it exerts backward. Machines rely on rotary motion. Wheels roll, screws turn and propellers spin. Even machinery without obvious rotary motion — looms, presses and typewriters — converts circular motion into linear motion. But nature turns no genuine wheels. The bodies of animals are laced with sinews, vessels and nerves which would rend and tear if their joints turned full circle.

Instead, animals' bones act as levers, moving from side to side and up and down but never revolving around a fixed axis. Yet, in locomotion, wheels and levers work in a similar way. A wheel, turning on its axle, is nothing more than a set of levers disguised as spokes, working one after another. Since what happens to the spoke once it leaves the ground is of no consequence, rolling wheels and paired legs function much the same. The major difference is that legs, swinging from the hips like pendulums, go forward when off the ground and backward when on the ground, while wheels turn only in one direction.

Muscles operate the levers formed of bones. In the arms and legs, each muscle is attached to two bones hinged by a joint. When a muscle shortens, the hinge swings open or shut depending on where the muscle lies. In taking a step, muscles on the back of the leg shorten to close the hinge at the knee, lifting the leg, as those at the front lengthen. Then, muscles on the front of the leg shorten to open the hinge at the knee, straightening the leg as it strikes the ground, while those at the back lengthen. Working against each other in

such antagonistic pairs, muscles drive the levers which move almost all large animals.

Different arrangements of bones and muscles help endow animals with the speed, agility, strength and stamina required by the places they inhabit and the ways in which they live. As dusk drops over a water hole on the Serengeti Plain in Tanzania, animals gather, displaying many ways of adapting to one environment. Herds of hooved animals — wildebeest, antelope and zebra — graze warily. Giraffes browse in the nearby stands of acacia. Skirting the herds are predators — cheetah, lion, wild dog and hyena — that prowl watchfully. At any moment a race, handicapped by centuries of evolution and run for the highest stakes, may break out on the plain.

The field is fast. All animals of the Serengeti are runners, or cursorials. But their speed comes from different sources and takes different forms. The long, lithe legs of hooved animals enable them to take long strides at an easy pace. They are built to cover long distances on the hot, dry prairie where water and grazing are scarce. The fastest among them, the wildebeest, zebra and smaller antelopes (especially the gazelle), carry relatively little weight on slender legs. Like fast-spinning wheels, their legs swing rapidly. This, together with a long stride and light body, gives them speed. With longer but heavier legs swinging more slowly and carrying more weight, the giraffe lopes at a gentler pace. All easy runners, hooved animals can cruise at high speeds over considerable distances. Consequently, the race for life is almost always decided near the start.

The predators stalk, getting as close as possible before attempting a kill. Only the wild dog and hyena have the stamina to stay with hooved prey for more than several hundred yards. Predators count on an explosive burst of speed to overtake their prey before it gets into stride.

Mounted on swift, nimble ponies, Plains Indians, right, picked their way among stampeding buffalo and rode close to place their arrows. For millennia, man has harnessed some animals to hunt others.

Stalking with stealth and striking with speed, this cheetah, left, has seized a Thomson's gazelle. For sheer straightaway speed, nothing outpaces the cheetah, fastest drag racer in the natural world.

The wild dog hunts more successfully, but nothing kills with the sudden speed of the cheetah. A bold, quick stalker, the cheetah steals to within fifty or sixty yards of its prey. Ready for the chase, it breaks into a trot. After a few steps, it bursts into a sprint. In two seconds, the cheetah covers sixty-five yards, hitting a speed of forty-five miles per hour. Another few seconds and it may top seventy miles per hour, enough to overtake any other animal. With long, light legs the cheetah takes lengthy strides, made even longer by the rippling of its spine, which flexes like a caterpillar. Unmatched for sheer acceleration, the cheetah only loses the chase when its prey, chiefly gazelle and impala, shake it with a sharp turn or confound it with an erratic flight. Quick and agile as they are, impala and gazelle fall to the cheetah on about 40 percent of its hunts.

On the Serengeti, prey and predator live in a balance of speed, but elsewhere the race is not always to the swift. Runners of tropical deserts and arctic tundra — camels and caribou — swap speed for stamina. Camels are slow. They cannot maintain their top speed, ten miles an hour, for more than a few strides, but they can cover thirty miles a day, carrying burdens of six hundred to a thousand pounds. Rather than tire themselves galloping, camels walk or pace. When pacing, they raise and advance both legs on one side at once, a rare gait in the wild. Because they burn food and lose water slowly, camels can last for weeks without water, even in the severe heat of the desert summer. Less hardy than camels, North American caribou migrate thousands of miles every year from taiga to tundra. By keeping to a trot and galloping only if pursued, caribou stay on schedule without tiring.

In thick forest, neither speed nor stamina can substitute for quickness and agility. To hunt and flee in dense vegetation, climbing is a useful skill. Many forest dwellers have short legs which, by lowering their center of gravity, lend them the stability required for arboreal living. Originally forest dwellers, baboons have extended their range and retained their agility as forests have shrunk. They feed on grass, shrubs and fruit, but will kill when the chance arises. Once they have tasted meat, they are likely to turn to hunting. Quick and agile, baboons capture hares, birds and young gazelle, often hunting in teams. Pound for pound the strongest of the big cats, leopards seldom chase. Instead, they stalk and ambush their prey, lunging at the last to kill. Leopards take impala and gazelle, among the fleetest afoot, then tote the carcass up a tree.

Prowess is not among the gifts of other animals. Slow and awkward, the elephant, hippopotamus, rhinoceros, bear and buffalo flourish, in part, because they are big. Most are vegetarians and none fear predators. They stand above the fray, waxing fat at the top of the food chain.

Without the speed of the cheetah, agility of the baboon or the bulk of the elephant, man masters and threatens them all. Although many animals are, in one respect or another, his physical superior, none adapts as successfully, hunts as proficiently or ranges as widely. All animals sense their physical limitations. They know

Athletes in action adorn much Greek art. Runners in armor race around this Athenian vase, begging the poet John Keats's questions "What men or gods are these? What mad pursuit?"

Artists have long tried to freeze motion as did the Roman artist who copied the now lost masterpiece Discobolus *by Greek sculptor Myron. Drawn taut with muscle, the marble exudes energy.*

when to pursue and when to flee, which to challenge and which to shun. They extend their range only when sure of survival. Man alone not only senses but overcomes his physical constraints, quickening the actions of his body with those of his mind.

A Nation of Athletes

Seldom have the two paced one another as closely as they did among the ancient Greeks. "In Greece," wrote the celebrated classicist Edith Hamilton, "man first realized what mankind was." The gods, cast in human form, appearance and character, were distinguished solely by their superhuman power. They embodied heroic ideals of honor achieved by action, and beauty, expressive of power, to which Greek mortals aspired. Man's body was his means to honor and his mirror of beauty, to be neither hidden nor flaunted but carried with confidence and pride. Friezes on temples, designs on urns and lines of poetry extol a beauty seldom of repose but alive with a power fired by the union of mind and body.

In games the Greeks sought their ideals. The historian Herodotus tells us that when Xerxes, king of Persia, was readying for war on the Greeks in 480 B.C., he found his enemy engaged in the games at Olympia. A companion, on learning that Greeks competed for nothing more than an olive branch, asked his king: "What kind of men are these that you have brought us to fight against — men who compete with one another for no material reward, but only for honor?" In fleeting action seen once and never repeated, the Greeks found the same beauty that moved an aspiring twentieth-century poet, the wife of a bush-league southpaw, to remark, "You can't rewrite a pitch." Athletic games in Greece were both ethical adventure and aesthetic experience.

In Homer's epics, generally thought to have been written in the ninth century B.C., games are common pastimes among heroic warriors. In the *Iliad,* a chariot race, boxing match, foot race, wrestling contest and javelin throw are part of funeral games for Patroclus, fallen friend to Achilles. In the *Odyssey,* after Odysseus feasts as the guest of the Phoenician king, young men entertain the company with games. Odysseus de-

murs at their challenge to compete, but when taunted takes up the discus and bests them all.

With the rise of the city-states in the eighth century B.C., games became a regular and prominent feature of social life. Called to arms at a moment's notice, citizens of warring city-states sought survival in fitness and prowess. "The Greeks," summed up classical historian Norman Gardiner, "were a nation of athletes." From early in childhood until well into manhood, Greek men exercised under trained eyes in the palaestra and gymnasium. The palaestra was a building, often privately owned by a schoolmaster, where young boys exercised. The gymnasium was an open field where professional athletes trained alongside ordinary citizens.

Despite their rivalries, the Greeks found unity in language and culture and celebrated it with festivals enlivened by athletic contests. Apart from countless local festivals, several national festivals pitted athletes of different cities against one another. Of these, the greatest was held at Olympia, foremost shrine of Zeus, whom the festival honored. Pindar, the fifth-century B.C. poet whose lyrics extolled many triumphs, tells us that Heracles, son of Zeus, founded the festival as a thanksgiving for victory in battle. At Olympia, as at other festivals, athletes performed to please the gods whose patronage assured victory. All festivals were times of truce, but none more than the Olympiad. Between 776 B.C. and 393 A.D., athletes, often from states at war, competed every four years.

The discus throw was added to the events in the *Iliad,* in early Olympiads. During the eighth century B.C., three running events — races of two hundred yards, four hundred yards and two or three miles — took the place of one. The pentathlon was also introduced. A contest of five events, the pentathlon included wrestling, running, jumping and javelin and discus throwing. The pankration, a form of hand-to-hand fighting, and horse racing were added a century later, along with several events for boys. Originally brief, the Olympic program grew to maturity by the sixth century B.C. With a race in armor put on the program in 520 B.C., the games were complete. Throughout Greece festivals followed the program and rules established at Olympia.

Honoring their gods, Greeks trained and competed naked. When the Olympiad began, athletes wore loincloths, only to doff them around 700 B.C. There is one story of a runner who lost his loincloth but won the race and another tale of a runner who fell to his death over a loincloth gone astray. Thucydides, who lived during the Peloponnesian War and recorded it, traces nudity in athletics to Spartans, whose success ensured emulation. The practice soon spread across Greece. To the Greek mind, shame at nudity marked a barbarian.

Sparta dominated the early Olympiads. Here, more than elsewhere, athletics were treated as preparation for war. Along with a demanding training regimen, Spartans underwent severe hardships to hone their bodies for competition. In Sparta, girls were encouraged to wrestle, run, throw, swim and ride so that they would bear strong, healthy children. Girls also competed regularly in separate festivals, and sometimes against boys. For more than 140 years, between their first victory in 720 B.C. and 576 B.C., Spartan diligence and self-discipline garnered more than half of the eighty-one recorded Olympic crowns. Today, we use the term "Spartan" to describe someone having the qualities of great courage, strength and austerity.

The rise of Rome and Carthage in the west and Macedonia and Thrace in the north eclipsed the supremacy of the original city-states. As Greek civilization spread around the Mediterranean, the popularity of athletics spread with it. But the more popular athletics became, the more competitions were staged for the entertainment of spectators rather than the pleasure of participants.

Pindar reminded athletes they must not seek to become gods. By the fourth century B.C., the Athenian soldier and historian Xenophon wrote angrily about the excessive adulation lavished on athletes, insisting that it was not "right to honor strength more than wisdom." More and more athletes competed for gain. As professionals swept the prizes, competition lost its appeal for the common citizenry. Training and competing solely for reward, athletes molded their bodies not to the ideals of beauty but to the demands of events. The original moral and aesthetic principles of Greek athletics steadily corroded.

Gaudy and Gory Spectacles

Romans despised Greek athletics. An active, martial people, the Romans thought exercise and recreation desirable and necessary. But competition in public, particularly naked, fell beneath their dignity and was best left to slaves and hirelings. When first-century emperor Nero offered a Greek festival, the historian Tacitus summed up the general ill will: "traditional morals . . . have been utterly ruined by this imported laxity! It makes everything potentially corrupting and corruptible flow into the capital — foreign influences demoralize our young men into shirkers, gymnasts and perverts." A sound Stoic fond of exercise, the writer Seneca frowned on professional athletes, saying, "It is foolish and quite unfitting for an educated man to spend all his time on acquiring bulging muscles, a thick neck and mighty lungs."

Like the Greeks, Romans honored their gods and fallen soldiers with games. Roman games were not contests but amusements. With the waning of the Roman Republic, the number and splendor of spectacles mounted. Vying with one another for the favor of a restless populace, politicians staged more diverse and extravagant shows. Chariot races at the Circus Maximus drew over a quarter of a million who clamored for the colors on which they wagered.

From Etruscans, a dying people of central Italy, Romans adopted combat between gladiators and fights with beasts. On the day in 249 A.D. that marked the thousandth anniversary of Rome, a thousand pairs of gladiators dueled in the sand of the Colosseum already soaked with the blood of more than 200 animals. Less grand but similar shows were held throughout the Roman Empire.

Fallen at the feet of his rival, a beaten gladiator, above, fears his fate at the imperial pleasure. Jean Gerome, a nineteenth-century French artist drew the capricious conclusion to the Roman spectacle. A Roman statue, The Pugilist, *right, expresses a violent trade. His face, as Simon and Garfunkel sang, "carries a reminder of every glove that laid him down or cut him till he cried out in his anger and his shame. I am leaving, I am leaving. But the fighter still remains."*

Citizens of Roman Antioch, once the capital of ancient Syria, responded to a Persian assault in 252 only when arrows rained on the race course.

Only in the eastern part of the Roman Empire, where Christianity originated, was athletic competition popular. There, perhaps because St. Paul, the apostle who won so many to Christianity, enjoyed athletics, they were tolerated and even encouraged by the church.

All spectacles embellished the pagan calendar. Most were violent, inviting condemnation, especially from Christians. A Carthaginian converted to Christianity, Tertullian had been schooled as a lawyer and attacked spectacles with the skill of a determined prosecutor. Wrestling was "the devil's own trade. . . . Its very movements are the snake's." Above all, he claimed the spectacles roused passions not fit for Christians. A crowd arriving at the Circus was "mad . . . disorderly, blind, excited already about its bet!" He damned all Roman excess, but his condemnation of public entertainments in *De Spectaculis,* written around the close of the first century, set the tone for later religious censure of games and pastimes by medieval churchmen and Puritan preachers alike.

The Ignoble Body

The exaggerated piety and harsh asceticism of the medieval church drew a veil of solemnity and sobriety over an earthy and ribald age. By the sixth century, Christian emperors had put an end to athletic festivals, gladiatorial combat and other spectacles. In villages across Europe, both nobles and commoners amused themselves with various games, pastimes and festivities, winking mischievously at the stern glare of the church.

More concerned for the flight of the soul than the grace of the body, medieval theologians were ambivalent about the benefits of physical exercise. Peter Abelard, the twelfth-century French

Earthy ecstasy and ribald revelry among dancers at village fetes, left, outraged pious churchmen. Pieter Brueghel the Younger, sixteenth-century Dutchman, painted their gaiety with unabashed frankness.

scholar better known for his tragic affair with Héloïse, persuaded himself that a sick body enlivened the mind. This was a view shared by his contemporary, Bernard of Clairvaux, who declared that "always in a robust and active body the mind lies more soft and more lukewarm." Others, equally pious and learned, disagreed. Robert Grosseteste, thirteenth-century chancellor of Oxford and Bishop of Lincoln, and Thomas Aquinas, the Angelic Doctor of medieval scholarship, endorsed Aristotle's ancient recommendation to exercise. But churchmen still looked askance at games.

Originating some time before the eleventh century, tournaments captured the fancy of the nobility. Decked with tapestries and trimmed with banners, galleries filled with spectators as knights, encased in polished armor and mounted on costumed horses, paraded in the lists. A free-for-all begun with the clash of thirty or forty

A stag sedately awaits the sting of a lady's sporting arrow in a medieval illustration. Archery, a popular pastime, was encouraged by kings as a means of keeping their subjects armed and ready for war.

Life in New York once moved at a slower pace when these early settlers enjoyed a game of bowls on the city's bowling green. This green is sometimes said to be the first sports field in the New World.

horsemen was the main event. Despite rudimentary rules, blood was shed and lives were lost. In jousting, knights wielded blunt lances, lessening the risks of injury and death. Other combative contests took the form of duels modeled on the tournament. Appalled at the bloodshed and revelry of tournaments, both papacy and crown sought vainly to ban them. Pope Alexander III withheld burial of the victims in 1179.

Commoners enjoyed less martial pursuits. Football, wrestling, bowling and hockey were popular games, often setting one village against another. With neither fast rules nor hard umpires, matches were apt to spark as much discord as geniality. The aftermath, rather than the game itself, drew the ire of the church. One young boy, Christopher Robson, was given "six yerkes with a byrchen rod on the buttocks" for kicking a football into the cathedral of York in 1466.

For the most part, civil authorities stood aloof. But in 1337, during the Hundred Years' War when Edward III was preparing England for war with France, he forbade all pastimes save archery on pain of death. Archery was always a popular pastime among the English and served the kingdom well at Crécy, Poitiers and Agincourt. In 1477, the House of Commons petitioned the crown that "no person should use any unlawful plays as dice, quoits, football and such like plays, but that every person mighty and able in body should use his bow, because that the defence of this land standeth much by archers." And in the mid-sixteenth century, Bishop Latimer, court preacher to Edward VI, praised archery as a "goodly Arte, a wholesome kinde of exercise, and much commended in Phisicke."

Apart from the violence and merrymaking accompanying games, clergymen reserved their strictest censures for dancing, which went hand in hand with popular festivities. Alluding to the custom in French villages, where a woman kept time with a bell, Cardinal Jacques de Vitry believed that just as a farmer kept track of his cows by tying bells around their necks, so the devil kept track of his souls. Throughout Europe, the church condemned dancing because it belonged to festivals having pagan roots, and it flaunted sexual license in the face of chaste convention.

During the fourteenth century, Italian humanists spurred European culture into motion. By enriching Christian theology with themes gleaned from classical writings, the humanists tempered the otherworldly strain in medieval culture. Scholar, diplomat and poet, Petrarch encouraged classical studies. Emboldened, others followed to challenge traditional scholarship and education. In place of dogmatic rationalism, the humanists sought to school the whole man — to bring mind and body into step.

Once merely tolerated, exercise and recreation began to be encouraged. Mapheus Vegius, a classical scholar who was close to the church, in a treatise on education recommended exercise to make students "more attentive and more eager" to avoid "the thing which is wont to ruin any good disposition: dejection and despondency." A contemporary, Vittorino da Feltre, required that his pupils play ball, fence, ride and hike in fair

weather and foul. He also devised a gymnastics program for adults. Aeneas Sylvius Piccolomini, later Pope Pius II, favored exercise for women as well as men and told the king of Bohemia that "games and exercises which develop the muscular activities and general carriage of the person should be encouraged by every Teacher." Many, among them the future pope, even approved of dancing. Those who did not said dancing dampened martial spirits.

From Italy, humanist studies spread north, reaching the towns of the Low Countries and the courts of England and France in the fifteenth century. There, as in Italy, humanists stressed proper schooling based on thorough study of the classics, but were slow to urge exercise and recreation. Erasmus, the Dutch humanist whose scriptural criticism helped fuel the Reformation, allowed exercise, but he still clung to the rift between the divine soul and the ignoble body.

But British author and diplomat of Henry VIII, Sir Thomas Elyot, prized exercise and recreation in the rearing of courtiers. In his *Boke Named the Governour,* he recommended virtually all exercise, pastimes and games, except football "wherein is nothinge but beastly furie and exstreme violence; wherof precedeth hurte, and consequently rancour and malice." Roger Ascham, once tutor to the young Elizabeth, recommended that a well-schooled courtly gentleman be able "to ride comely, to run fair at the tilt or ring; to play at all weapons, to shoot fair in bow or surely in gun; to vault lustily, to run, to leap, to wrestle, to swim; to dance comely, to sing, and play on instruments cunningly; to hawk, to hunt; to play at tennis, and all pastimes generally."

Henry VIII enjoyed tennis on the splendid court built in his palace at Hampton. A century later, Louis XIV built a tennis court at Versailles on which, in 1789, the Estates-General served no-

The modest costumes and boyish exuberance of nineteenth-century French artist Henri Rousseau's Football Players *contrasts sharply with the harsh determination of today's professional game.*

tice of revolution. Of a child's education, French essayist Michel de Montaigne insisted, "It is not enough to harden his mind; we must also toughen his muscles."

The benefits of exercise and tolerance of recreation were soon challenged by the religious zeal of radical Protestant reformers. Martin Luther favored exercise and John Knox once came upon John Calvin bowling, on a Sunday. But Puritan sects, harking back to the Middle Ages, condemned traditional festivals for their heathen origins, jealously reserved Sundays for holy worship and, of course, denounced dancing.

Defenders of tradition rested their case on the antiquity of customs linked to the rhythm of seasons. Puritans believed custom counted for nothing unless sanctioned by the word of God. In Dorset, England, William Kethe preached, "The Lord God hath commaunded . . . that the Sabbath day should be kept holy," but "the multitude call it there revelyng day, whiche day is spent in bulbeatings, bearebeatings, bowlings, dicying, cardyng, daunsynges, drunkennes, and whoredome." To the Puritan conscience, recreation and sacrilege were one. And nothing outraged Puritans more than dancing. John Northbrooke, also a preacher, fumed, "They daunce with disordinate gestures, and with monstrous thumping of the feete, to pleasant soundes, to wanton songues, to dishonest verses. Maidens and matrones are groped and handled with unchaste hands, and kissed and dishonestly embraced: the things which nature hath hidden, and modestie covered, are then oftentimes by meanes of lasciviousnesse made naked, and ribauldrie under the colour of pastime is dissembled."

In England, James I took up the Puritan challenge in 1618, issuing a declaration on sports. "When shall the common people have leave to exercise if not upon the Sundayes and Holidays, seeing they must apply their labour, and winne their living in all working days?" Protests grew more shrill, and when Puritans came to power on the point of Oliver Cromwell's sword, the declaration was burned by the hangman.

Ignoring James's challenge, the New England Puritans tightened restrictions on play. On Christmas Day, 1621, Governor Bradford was dismayed to find some newcomers to Plymouth "in the streete at play, openly; some pitching the barr and some at stoole-ball, and such-like sports." Deprived of recreation, more and more colonists sought out the tavern. As those who did not share the faith of the founders flocked to the New England colonies in rising numbers, Puritan austerity became increasingly difficult to enforce. The bough bent before it broke as prohibitions against popular entertainment were lifted. But the Puritan conscience cast a long shadow over the history of New England.

In England, the political edifice of Puritanism ended with the fall of the Commonwealth, but many of its attitudes toward work and play survived. A cardinal virtue among Puritans, industry continued to be prized as workshops and factories spread during the eighteenth century. The factory system required a steady, disciplined regimen of labor, which many thought incompatible with traditional recreation. To harness working people to the new machines, entrepreneurs urged the government to restrict holidays and festivals. Josiah Wedgwood, the famous pottery manufacturer, complained that his labor force put play before work, making merry when he expected them to make pots.

To the chorus of industrialists, the Evangelicals added a refrain reminiscent of their Puritan forerunners. As one inspired poet rhymed:

> In Works of Labor or of Skill
> I would be busy too:
> For Satan finds some mischief still
> For idle Hands to do.

Both inside and outside the established church, the evangelical movement had a profound impact on English society, contributing to standards of sobriety and propriety demanded by a country more and more ruled by balance sheets, humming machinery and factory bells.

Exercising the Body Politic

While the popular entertainments ran afoul of industrial discipline and religious enthusiasm, exercise won favor during the Enlightenment of the eighteenth century. Regarding themselves as heirs to the Greeks, the philosophes enhanced

Machinery spared man's muscles but paced his life with its relentless rhythms. As men moved from field to factory, leisure time grew more structured and recreation became more formal.

their legacy. In writings on education, the union of mind and body was a persistent theme. Easily the most radical spirit of the age, Jean Jacques Rousseau thought the education of youth a powerful instrument of reform. In *Emile,* he prescribed an ideal education. "To learn to think," Rousseau argued, "we must . . . exercise our limbs, our senses, and our bodily organs, which are the tools of the intellect." An admirer of Sparta, Rousseau thought physical and moral strength contributed to each other. "The great secret of education," he believed, "is to make the exercises of the body and of the mind always serve as a recreation for each other."

His ideas were warmly received in Germany where schools embodying them were established. At one, the Schnepfenthal Educational Institute, teacher Johann Friedrich Guts' Muths became perhaps the first authority on physical education. In his *Gymnastics for the Young* and *Games,* written in the 1790s, Guts' Muths explored the strengths and skills developed by various forms of exercise. Pioneered in some schools, physical training took a major role in German public life. Defeated and humbled by Napoleon in 1806, the German states responded with a movement for national unification and political liberalism. Near the forefront was Friedrich Ludwig Jahn, who championed gymnastics as a means to the physical and moral revival of German nationality. A schoolteacher, Jahn opened the first *Turnplatz,* or playground, outside Berlin in 1811. Modeled on the Greek palaestra, the *Turnplatz* was equipped for various exercises and games, though the emphasis fell on synchronized gymnastic displays. At first Jahn catered to schoolchildren, but soon he arranged meetings for adults as well. Frankly political in purpose, the *Turnverein,* Turner Societies, contributed to the resurgence of Prussian power which helped topple Napoleon in 1815. By 1819, however, they threatened the established order and were outlawed.

When the ban lifted in 1840, the *Turnverein* flourished again, this time helping to kindle the revolutions of 1848. The wave of repression that followed washed many members to America's shores where, in 1850, they founded the National Turnerbund. By the Civil War, 150 Turner societies with ten thousand members had sprung up in German communities across the United States. Meanwhile, in Germany, gymnastic enthusiasts jettisoned political aspirations and gained official approval. Societies grew and festivals flourished. In 1913, a quarter of a million people watched sixty thousand gymnasts perform at Leipzig.

Gymnastics, again serving patriotic purposes, also attracted a large following in Scandinavia. In the early decades of the nineteenth century, Denmark became the first country to introduce compulsory physical education in its schools. Team games captured the English fancy much as gymnastic exercises appealed on the Continent.

Team sports have long been enjoyed. Ancient and primitive peoples have played a variety of such games. A hundred years after Christ, the Mayans of Central America played a ball game in a walled court, forcing the ball into a raised goal without using their hands. Other forms of football and polo were played in ancient China. North American Indian tribes played lacrosse, sometimes for days. Little is known of these games, but authorities agree that most were more ritual than sport. Team sport, played to rules ensuring fair play, originated in the great schools of England during the nineteenth century.

Like martial marionettes, German soldiers swing and sway, teeter and totter on gymnastic apparatus. In the Napoleonic era, such gymnastic exercise and performances helped to kindle German patriotism.

Schoolboys first tailored their games to the peculiar features of school buildings and grounds. Fives, a forerunner of handball, was played at Westminster against the buttresses of the abbey walls. Football at Tonbridge used the gutters ringing the gravel courtyard. At Eton, the "wall game," an arcane kind of rugby, moved back and forth between a garden wall and a venerable tree with more gusto than goals. Often, boys of a single boarding house took on the remainder of the school. Rules were lax, mercy rare and brutality common. A keen player of the "wall game," renowned economist Sir John Maynard Keynes wrote his father in 1902 that it was "impossible to make the Wall Game humane and one suitable for the newly shaved and tender . . . the present form of legalised ruffianism is the best condition under which one can play the glorious game."

The Principle of Fair Play

These parochial pastimes gave way to formal games between schools during the nineteenth century. Cricket, rugby and football were the staples of English sport. Eton and Westminster had played the first recorded cricket match in 1796, but regular competition began after 1800. In 1806, Eton first met Harrow with the young romantic poet Byron scoring poorly for an outclassed Harrow. He then got drunk after the match. Rugby developed at the school from which it took its name, under the reforming zeal of headmaster Thomas Arnold. Father of the poet Matthew, Thomas Arnold had found the school riddled with abuses and lacking in spirit. Along with intellectual and disciplinary rigor, he encouraged games to instill moral purpose and manly character in the boys of Rugby. As his success at Rugby was emulated by other schools, games came to consume the life and passion of English schoolboys.

Popular by 1850, games soon became an obsession. The headmaster of Sherbourne offered a thanksgiving sermon for victories ensured, in part, by his own fervent cheering in English, Greek and Latin. Exaggerated claims for the role of games in the prosperity, stability and expansion of Victorian Britain were seldom disputed. The most celebrated arose when, in 1855, a Frenchman writing a biography of Wellington chanced upon boys at games in the fields of Eton. Imagining Wellington revisiting the scene, the writer had him remark: "It was here that the battle of Waterloo was won." Wellington had actually loathed his two years at Eton, avoiding

Coal, mined by muscles, fired the Industrial Revolution. Women work alongside pit ponies in Leonard Defrance's scene of an eighteenth-century mine shaft. Machines have replaced ponies, but not manpower.

games and preferring a tranquil garden near his boarding house. Most likely it was Napoleon, not Wellington, who made the famous remark. Games did less to expand the British Empire than the British did to spread the popularity of games.

The British games changed once they were adopted elsewhere. In America, cricket turned into baseball and rugby became a unique brand of football. As for football itself, it spread around the world as soccer, arousing more fervor and enthusiasm than any other sport. Although the games changed, the principle of fair play, ensured by precise rules, became Britain's greatest legacy to sport.

Mastering Mechanical Energy

Man moves most gracefully at play, but most purposefully at work. Other animals use natural objects as tools and some even fashion tools. But man alone designs and makes them. By enhancing the power of his muscles and capturing powers beyond them, this ability marks him as a methodical manipulator of energy. Until well into this century, in all societies, much of the energy to work tools and drive machines was generated by the muscles of men and animals.

In seeking to expand the power of muscles, man mastered mechanical energy. He handled his earliest tools and drove his first machines — plows, sledges and travois — with his own muscle. Then, perhaps in the fourth millennium B.C., he hitched draft animals, particularly oxen, to his plow. By harnessing animals, man pulled himself across a threshold. For the use of animal power led to the invention, or at least widespread adoption, of the wheel, the essential component of all machinery. Wheeled vehicles and rotary machines, pulled and driven by men or animals, set the pace of progress until the advent of steam in the eighteenth century.

A yoke of oxen turn soil in India as they have done for thousands of years. Working as tractors, animals till some three-fourths of the world's arable lands, from wheat fields to rice paddies.

Steam did not always best muscle easily. In the 1870s, the Chesapeake and Ohio Railroad was cutting through the mountains of West Virginia at Big Bend Tunnel, then, at a mile and a quarter, the longest railway tunnel in the world. Men in teams of two drilled the rock. The shaker straddled a six-foot steel drill, twisting it a quarter turn each time the driver, swinging a ten pound hammer through a nineteen foot arc, struck it. At Big Bend, no one drove steel like John Henry. His power rivaled that of the newly invented steam drill. Swinging a twenty-pound hammer in each hand, John Henry drilled two seven-foot holes in the thirty-five minutes it took the steam drill to make one nine-foot hole. And, despite the song, he did not die with his hammer in his hand. Men may no longer row galleys or drive steel, but on docks, roads, buildings and railways in even the most technologically advanced countries, muscle still performs as machine.

Working animals are even more common. Yaks plod on the cold, high plateaus of the Himalayas and camels trek across the hot, dry deserts of Arabia. Three-fourths of the world's agricultural land is tilled with the help of animals. The horse, one of the more efficient traction animals, plows the plots of peasants in eastern Europe. Buffaloes work the swampy ground of southern Asia and the Nile valley. In many parts of Asia, elephants account for most of the power in logging operations. The camel runs the stationary machinery of mills in India, the Sudan and the dry parts of Africa and, along with the buffalo, turns various devices to raise and spread water.

Just as man devised machinery from the application of muscle power, he also learned more about muscles from principles of mechanics. The study of body motion, formally kinesiology or biomechanics, originated — and still proceeds — from mechanical principles.

Leonardo's careful dissections and practiced eye enabled him to picture musculature with unprecedented skill. These drawings of the arm show muscles and bones working together in movement.

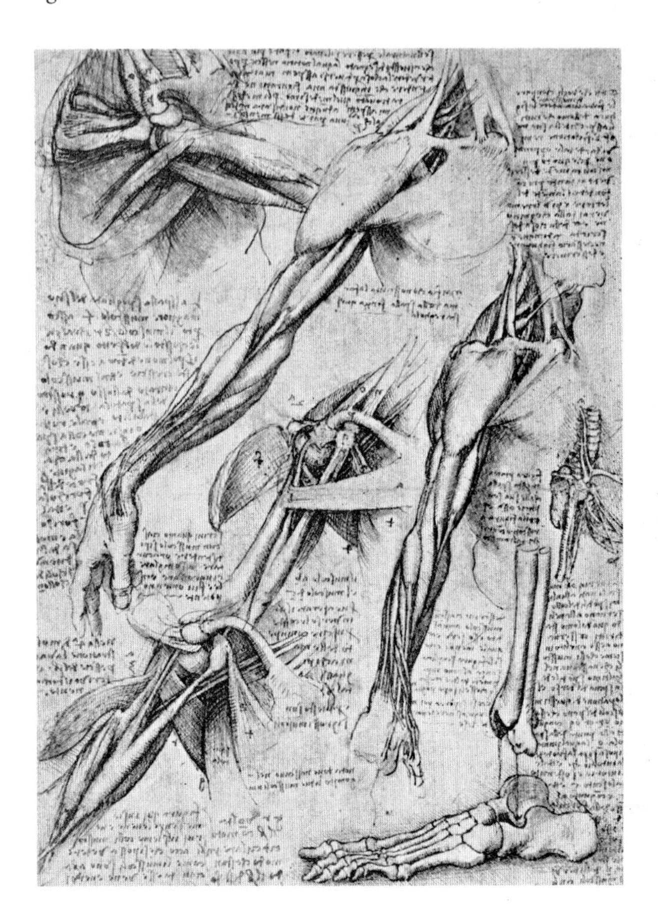

Although he had performed no experiments, Aristotle deduced fundamental principles of animal motion twenty-four centuries ago. In three treatises — *Parts of Animals, Movement of Animals* and *Progression of Animals* — he framed laws of motion, first describing the action of limbs and then reducing the motion to geometric principles. Watching men move, Aristotle grasped that to walk erect an animal's legs must shift its weight around its center of gravity. By appreciating the significance of the center of gravity, he was able to explain why man's legs are stronger than his arms and why children, with upper bodies that are disproportionately heavy, master walking as their stature shifts. He also understood how jointed limbs serve motion. He distinguished between flexion, the change from a line to an angle, and straightening (now called extension), the change from an angle to a line. When walking, he concluded, if one leg were "thrust forward, the other must of necessity be bent." By analyzing the operation of the legs as hinged levers, he explained how the rotary motion of bones about joints turned to translatory motion, or movement in a straight line. Later refined and elaborated, Aristotle's findings marked the starting point for the study of motion.

Some five hundred years later in Rome, Galen probed within the body to describe the engines that made it move. The pioneer of anatomy who learned by dissection, he explored the positions and actions of muscles. He explained how they worked in opposing pairs, agonists and antagonists, to move the bones at joints. His knowledge of anatomy enabled him to distinguish between motor and sensory nerves, dropping a vital clue to the source of movement. The first to realize that muscles operated by contracting, he believed that "animal spirits," coursing along the motor nerves from the brain, caused the muscles to shorten. Along with Aristotle's mechanical principles, Galen's anatomical inquiries described the bounds of kinesiology.

During the Middle Ages, the literal lessons, if not the inquisitive spirit, of classical learning were sanctified by the church as authentic knowledge. Dissection was discouraged, even forbidden, except for sporadic exhibitions using lower animals. Until the fifteenth century, the study of movement stood still.

It was Leonardo da Vinci who picked up the pace. "Motion," Leonardo scrawled in his notebooks, "is the cause of all life." He turned his mind and eye to solving its mysteries. An ingenious engineer, Leonardo tested mechanical principles with structures and machines of his own design and construction. An exacting artist, he studied anatomy and physiology, guiding his pen and brush with knowledge of the form and working of the body. Armed with mechanics and anatomy, Leonardo stormed the body for science.

By observing the poise of the body, he sensed that its center of gravity must remain at or near the center of its base of support. In the *Treatise on Painting*, he noted that when standing on one leg, a person must extend a limb or a hip to maintain his balance. But, to sit up or move ahead, the body's weight must shift forward

Leonardo da Vinci

Making Art of Science

"The eye, the window of the soul," said Leonardo da Vinci, "is the chief means whereby the understanding can most fully and abundantly appreciate the infinite works of Nature." A man of diverse talents commanded by a singular genius, Leonardo observed natural phenomena with the precision of a scientist, the insight of an artist and the curiosity of a child.

By the age of eighteen or twenty, Leonardo was apprenticed to Andrea del Verrocchio in whose Florentine workshop he studied painting, sculpture and mechanics for seven years. When he left in 1477, he had fulfilled his own saying: "Poor is the pupil who does not surpass his master." Already an accomplished artist, he began to pursue scientific inquiries.

Convinced that Nature revealed her secrets to perception and analysis, Leonardo confined his studies to phenomena that could be experimentally tested and mathematically measured. He distinguished clearly between art and science. This master of both insisted the sciences were "not concerned with quality, the beauty of Nature's creations." He turned to traditional authorities as guides, only to abandon them when his eyes led him elsewhere.

He studied anatomy as an apprentice, perhaps even by dissection. Many Renaissance artists, tending to picture the body realistically, witnessed and performed dissections. But by 1487, Leonardo had embarked on an exploration of the body far more searching than his peers dared or art required. Wanting not only to depict the body but to understand it, he endured many nights "in the company of corpses, quartered and flayed, and horrible to behold." In Leonardo's dissection of a pregnant woman, art historian Sir Kenneth Clark glimpsed "noble and passionate curiosity" which drove a tender and delicate man "to make such a terrible dissection, and to draw it with such a lucid and purposeful touch."

Leonardo's knowledge of anatomy and mathematics inspired his studies of muscles. He distinguished many different muscles by their shape, size and movement, and noted their arrangement with tendons. He followed the course of arteries, veins and nerves through muscles, realizing that "muscles retract or extend solely on account of the nerve from which they receive a stimulus." Although he drew muscles accurately and beautifully, their action fascinated him more than their form.

To observe the operation of muscles, Leonardo constructed a model. He attached cords and wires, representing muscles, to a skeleton. He saw that when one of a pair shortened, its partner lengthened. Moving from model to easel, Leonardo diagrammed the mechanical forces generated by muscles to lever the skeleton. In detailed drawings, he showed how the configuration of muscles changed as they moved limbs.

What Leonardo drew overshadows what he wrote. He teaches as he learned, by the eye, inviting others to share his rare vision of nature.

By discovering the force of gravity and framing universal laws of motion, Sir Isaac Newton steered the planets around their orbits. Poet and painter William Blake, in 1795, drew Newton as a herculean figure to suggest his power of mind in his strength of body. Crouched over a compass, Newton seems to be placing the planets in their paths.

from its center of gravity. This shift, he said, is involuntary, as when a man leans forward to mount a steep slope.

His knowledge of anatomy enabled him to reconstruct the operations of muscles working to move the body. He animated a skeleton by attaching copper wires where muscles should be. Pulling the strings of his puppet, Leonardo saw clearly that, because muscles were paired, one extended as the other contracted. Many of his careful drawings show how paired muscles work together to produce different movements of limbs. Whether playful or deliberate in design, Leonardo took pains to make his notebooks unintelligible to unauthorized readers. Published after his death, they reveal a mind that learned infinitely more itself than it taught to others.

In 1543, twenty-four years after Leonardo's death, Polish astronomer Nicholas Copernicus lay on his deathbed, holding one of the first printed copies of his work, *On the Revolution of the Heavenly Spheres.* By placing the sun at the center of the universe, Copernicus sowed doubt and confusion over the whereabouts and movements of the planets. "And new philosophy calls all in doubt," penned English poet John Donne, who went on to lament:

> We think the heavens enjoy their spherical,
> Their round proportion embracing all.
> But yet their various and perplexed course,
> Observed in divers ages, doth enforce
> Men to find out so many eccentric parts,
> Such divers down-right lines, such overthwarts,
> As disproportion that pure form. It tears
> The firmament in eight and forty shares.

Accustomed to knowing with certainty where to find heaven and hell, men were at once distressed and fascinated to lose themselves in the celestial traffic. The scientific revolution that followed flowed from efforts to explain the motions of heavenly bodies. In so doing, science established principles and developed methods that also accounted for the movements of human ones.

Astronomer Galileo believed "the language of Nature" was "written in the symbols of mathematics." Through his telescope he eyed the stars, but Nature spoke to him at his desk where he

Carefully arranged pairs of legs line Luigi Galvani's laboratory table. In 1786, the Italian scientist found that leg muscles of dissected animals twitched in response to electrical stimulation from his scalpel.

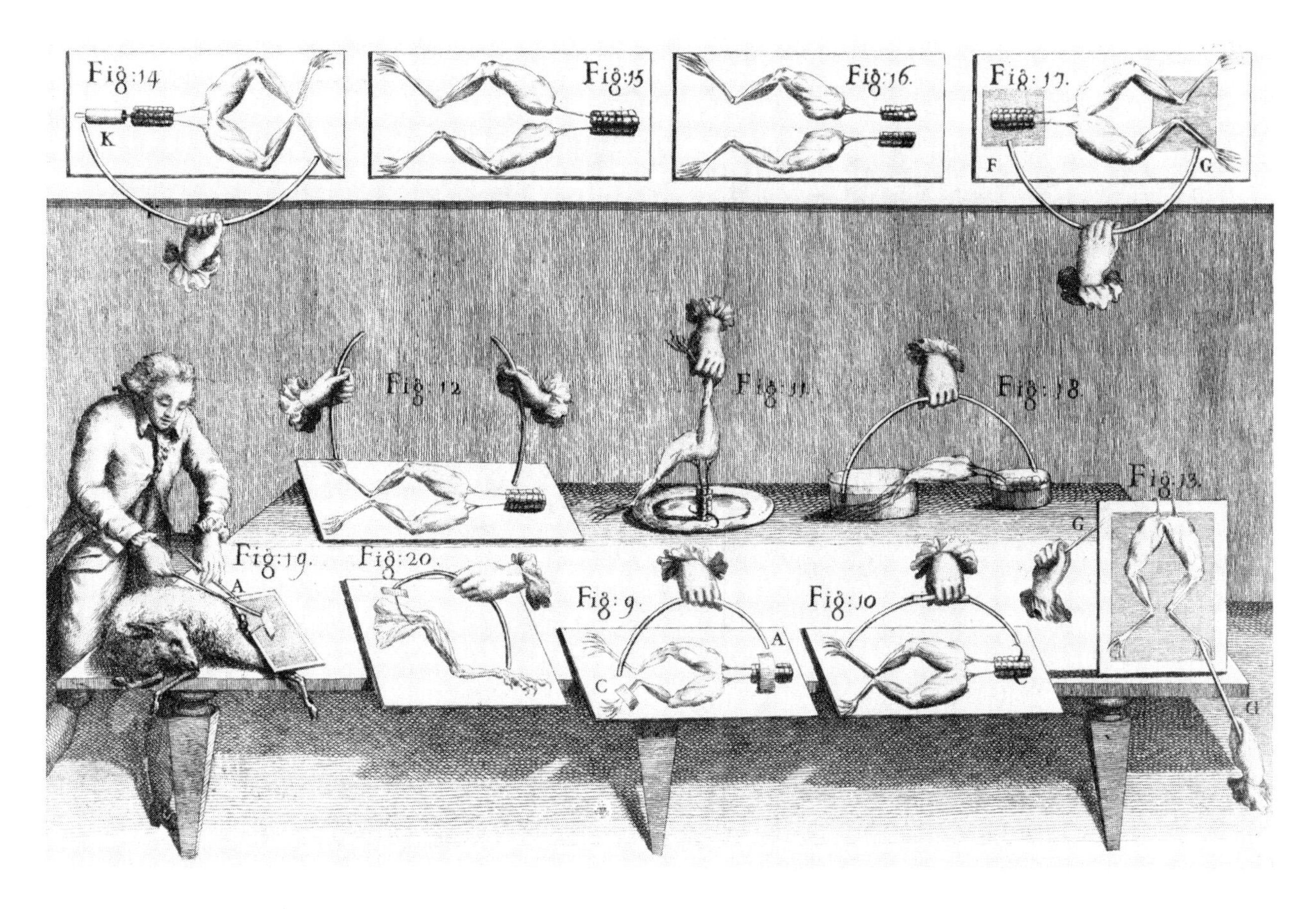

made his greatest discoveries. With axioms and theorems, equations and functions, he plotted the motions of falling bodies, rolling balls, swinging pendulums and arching shot. He measured the structure and movement of man with similar yardsticks, but only fragments of his manuscripts survive. He confined science to measurable phenomena — weight, mass, acceleration, velocity, space and time. His reckonings marked the beginning of classical mechanics, a treasure that gathered interest with the efforts of others.

A student of one of Galileo's pupils, Alfonso Borelli was an accomplished mathematician and physicist who, mixing with anatomists and physiologists at Pisa, turned his hand to the human body. His book, *On the Motion of Animals,* published in 1630, treated animals as machines moved by levers which were, in turn, powered by muscles. Borelli sought to measure forces generated by different muscles as well as the loss of power caused by malfunctions and resistance. Yet he also realized that mechanical actions were fueled by complex chemical processes which caused muscles to contract. But, like Galen long before him, he fastened on the flow of "animal spirits," stimulated by the brain, to account for contraction. Filling and enlarging pores in the muscles, the spirits reacted with an unknown substance, sparking something akin to fermentation to bring about contraction.

Borelli's theory of muscle contraction was quickly challenged by Francis Glisson, an English physician best known for his work on rickets. By experiment, Glisson showed that muscle fibers, far from expanding under the pressure of spirits, contracted. "The idea of the explosion of spirits and the inflation of muscles has now been silenced," Glisson confidently concluded. The foremost physiologist of the eighteenth century, Albrecht von Haller elaborated Glisson's find-

Knowledge of motion leapt forward with the invention of moving pictures. With techniques developed by Frenchman Etienne Jules Marey, American artist Thomas Eakins photographed this boy jumping.

ings, proposing that the capacity to contract was an innate quality of muscles, quite independent of the nervous system. Haller's conclusions were questioned in turn by a contemporary Scottish physician, Robert Whytt. In *An Essay on the Vital and Other Involuntary Motions of Animals* of 1751, Whytt argued that movement sprang from an unconscious sentient principle mediated through the spinal cord.

Confusion over the control of muscle contraction may have arisen from the different actions of different muscles. In 1700, Italian physician Giorgio Baglivi distinguished smooth muscles, built for sustained work, from striated muscles, those geared for sudden actions. Nicolaus Steno, Danish anatomist, described the heart as a simple muscle, perpetually working without a visible hand on the pump handle. Obviously, the nervous system exercised different kinds of control over different kinds of muscles.

During the late eighteenth century, studies of the electrical stimulation of muscles suggested the means, if not the pattern, by which nerves governed muscles. Luigi Galvani noticed that when touched with a scalpel, the muscles of a dissected frog contracted. He concluded that an "indwelling electricity . . . proceeded along the nerve" and he then sought a way to measure it. Electrical current took the place of animal spirits, sparking intensive research among physiologists. In the next century, Guillaume Duchenne pioneered the use of electricity to treat disease. His *Physiologie des mouvements* of 1865 thoroughly catalogued the functions of individual muscles as they moved the parts of the body.

Despite advances in mechanics and dynamics, the study of animal movement languished for want of a way to reproduce the chronological sequence of the phases of motion. Since antiquity, man has sought to capture motion in his art.

Eadweard Muybridge

Stepping Into the Frame

Learning springs from seeing, either seeing something unknown for the first time or seeing something familiar in a fresh light. Nothing is more commonplace than the movements of people and animals. Yet neither were closely observed and clearly understood until Eadweard Muybridge's cameras first matched motion stride for stride a century ago.

Coming to California from England, Muybridge earned fame as Helios, the Flying Camera, for his landscapes of the Yosemite Valley, shot early in the 1870s. He turned from still lifes to moving images when Leland Stanford, a wealthy horse breeder, engaged him to discover whether galloping horses lifted all four of their hooves at once.

No sooner had Muybridge begun to master the technical problems than he was interrupted. First he was called to photograph the Modoc Indian War which broke out in northern California in 1873. Then, he found himself caught up in a tragic triangle. He returned to learn that his wife had been unfaithful and was carrying a child. In the fall of 1874, he was charged with the fatal shooting of his wife's lover. After a taxing trial, Muybridge was acquitted and departed for Latin America in 1875, com-

missioned by Stanford to photograph "all the curious places a traveller can see." His pictures reflected his growing technical skill and were highly acclaimed. When he returned to California two years later, he finally aimed his camera at the gaits of horses.

The next year Muybridge successfully photographed running horses at Stanford's track in Palo Alto, freezing the action from frame to frame to stop their steps. His photographs, no less than the equipment and techniques he used to take them, stirred a sensation. Within two years, he devised a means of projecting continuous images onto a screen to become the first to show moving pictures.

Now celebrated around the world, Muybridge set off on a lecture tour of America and Europe. In Paris, painters and sculptors who sought to capture in oil and stone what he seized on film were thrilled by his work. There he met Etienne Jules Marey, the photographer whose early studies of motion had aroused Leland Stanford's original interest.

On returning from Europe, Muybridge left California for Philadelphia, where, with the patronage of J. B. Lipincott, he embarked on the close study of human and animal movement. Designing his settings for artists and scientists, he took 100,000 photographs between 1884 and 1886. With a whimsy betraying his rather eccentric character, Muybridge put his nude models through some peculiar paces. He shot one girl spilling a bucket of water over another as well as a mother spanking her child. At the zoo he recorded the movements of animals in a series of studies, many of them never repeated. Later published in two volumes, prefaced by his own analyses of gaits and movements, Muybridge's photographs enabled all to see what before they had merely glimpsed.

Until Eadweard Muybridge began to photograph moving animals in the 1870s, right, the choreography of gaits eluded even the sharp eyes of artists. At left, a neolithic painter doubled the number of a boar's legs — a noble effort at picturing motion, frozen forever in the depths of a Spanish cave. Gaits of horses proved particularly puzzling. The Currier & Ives print, below, puts a high-stepping trotter through its correct, natural paces.

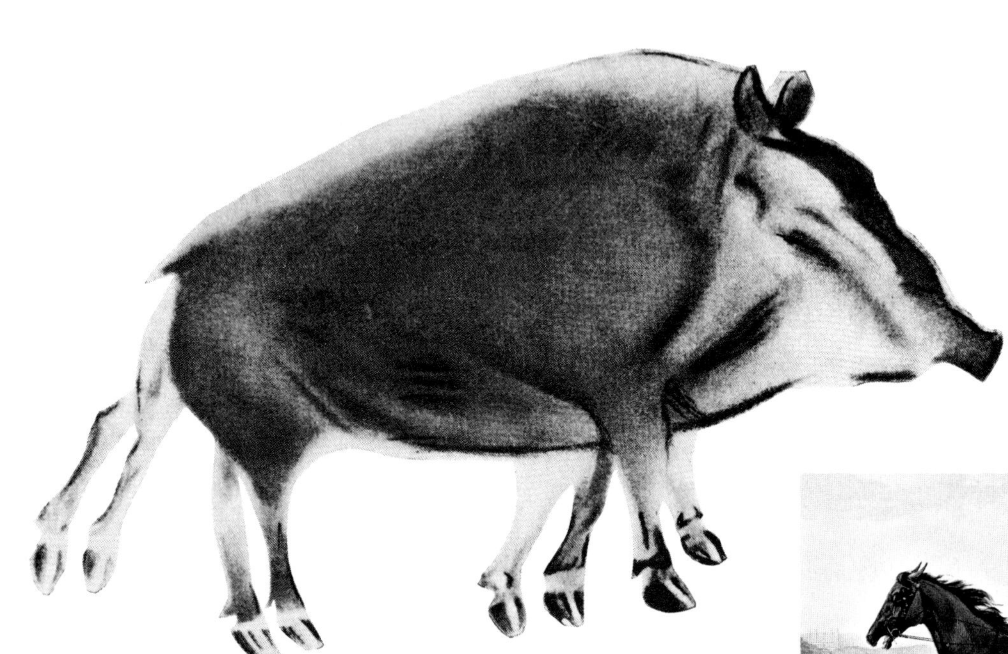

Primitive cave painters added extra legs to animals to suggest movement. Renaissance artists deliberately distorted form to express action. Though filled with beauty, the eye of the keenest artist was not fast enough to seize the rapid, subtle movement of flashing hooves and beating wings. Until well into the nineteenth century, perhaps only a poet, Emily Dickinson, could slow a hummingbird in flight:

> A Route of Evanescence
> With a revolving Wheel —
> A Resonance of Emerald —
> A Rush of Cochineal —
> And every Blossom on the Bush
> Adjusts its tumbled Head.

One afternoon in 1872, Leland Stanford, the former governor of California turned railroad magnate, was discussing horses' gaits with a friend. Both men owned, bred and raced blooded stock. And both were familiar with an article by Etienne Jules Marey, professor of medicine at Paris. Marey claimed to have demonstrated by experiments that all four of a horse's hooves sometimes leave the ground. Neither Stanford nor his friend was convinced, however. Stanford, persuaded to engage a photographer to find out if Marey was right, hired Eadweard Muybridge. An Englishman drawn to California by the Gold Rush, Muybridge had become the most famous photographer on the West Coast.

Muybridge pursued the enterprise with skill and perseverance, overcoming formidable technical obstacles to ensure reliable results. He experimented with different techniques until 1878, when he and Stanford, satisfied with the results, held a public demonstration at Palo Alto. Twelve cameras were mounted alongside the track, each triggered by the motion of the horse as it ran past. As a trotter ran the course, heavily sprung

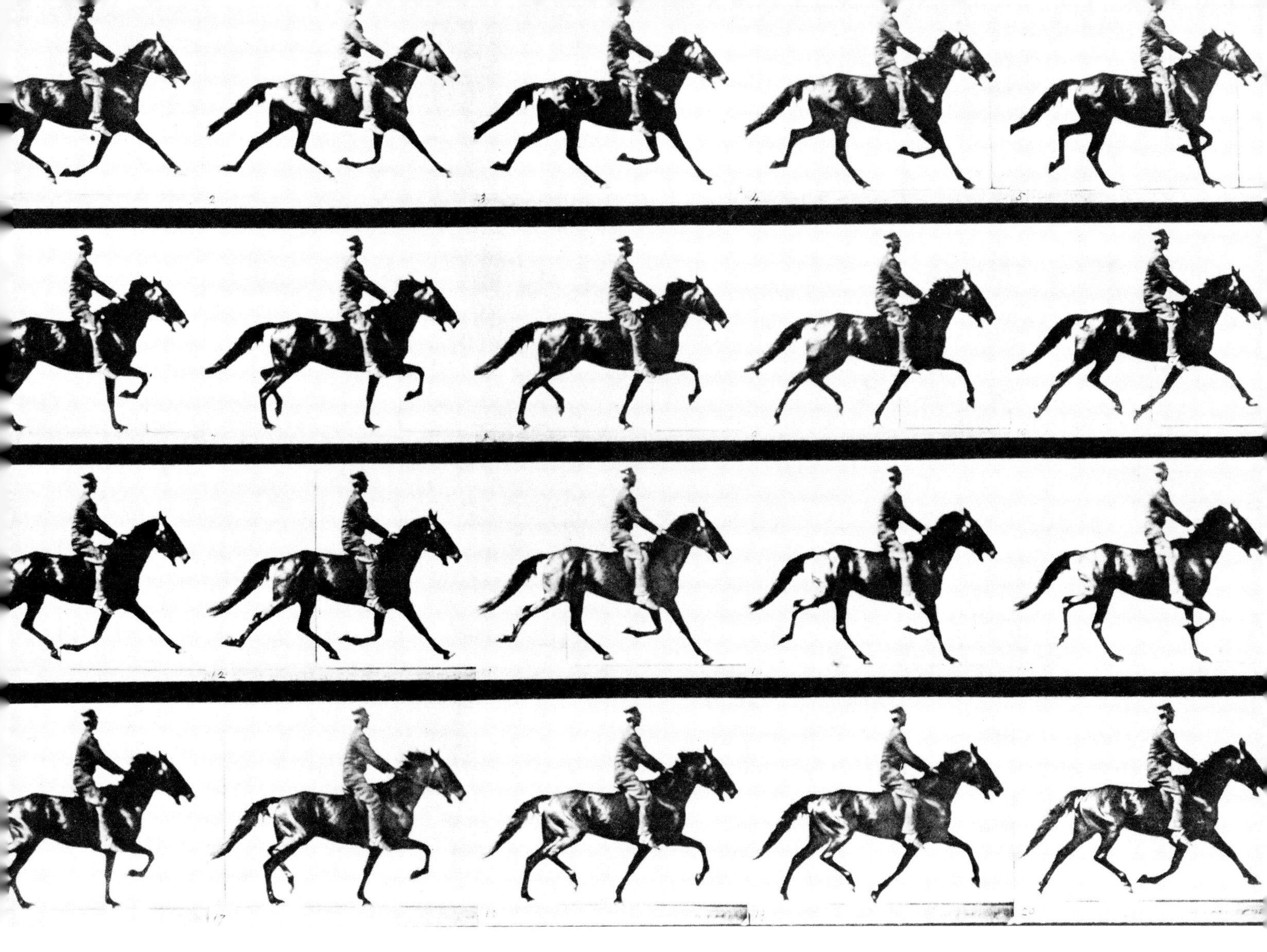

shutters opened and closed in a flash. Crowding into the nearby darkroom, the press was astonished to witness "the true story of the stride of a horse." When Kentucky mare Sallie Gardner galloped by, snapping her saddle girth, the camera caught every detail, including her four hooves in the air at once. Muybridge's genius solved a puzzle that had bewildered man for millennia. He had invented a powerful instrument for observing animal motion.

Meanwhile, knowledge of how muscles generate motion grew apace. Using electrical stimulation techniques, American Henry Pickering Bowditch proposed the "all-or-none" principle of muscle contraction, which holds that if a muscle fiber contracts at all, then it contracts as much as it can. The strength of the stimulus determines the number of fibers called on to contract. By observing single muscles and muscle groups at work, English neurophysiologist Charles Edward Beevor refined much of the thinking on contraction. Critical of relying exclusively on electrical stimulation, Beevor argued that the technique demonstrated only what muscles may do, not necessarily what they do do.

Beevor's suspicions were confirmed by the work of fellow Englishman Charles Sherrington, who published *The Integrative Action of the Nervous System* in 1906, which stands as a landmark in the literature of kinesiology. Sherrington observed that the neural impulses stimulating the nerves of a particular muscle also inhibit the nerves of its antagonist. Known as the principle of reciprocal innervation, this phenomenon ensures the harmonious working of muscles in pairs and groups. Like others, Sherrington agreed that man was a machine, but one with a mind. Just as Sherrington's work highlighted the harmony of the body's muscles, so he prized the classical belief in the union of mind and body.

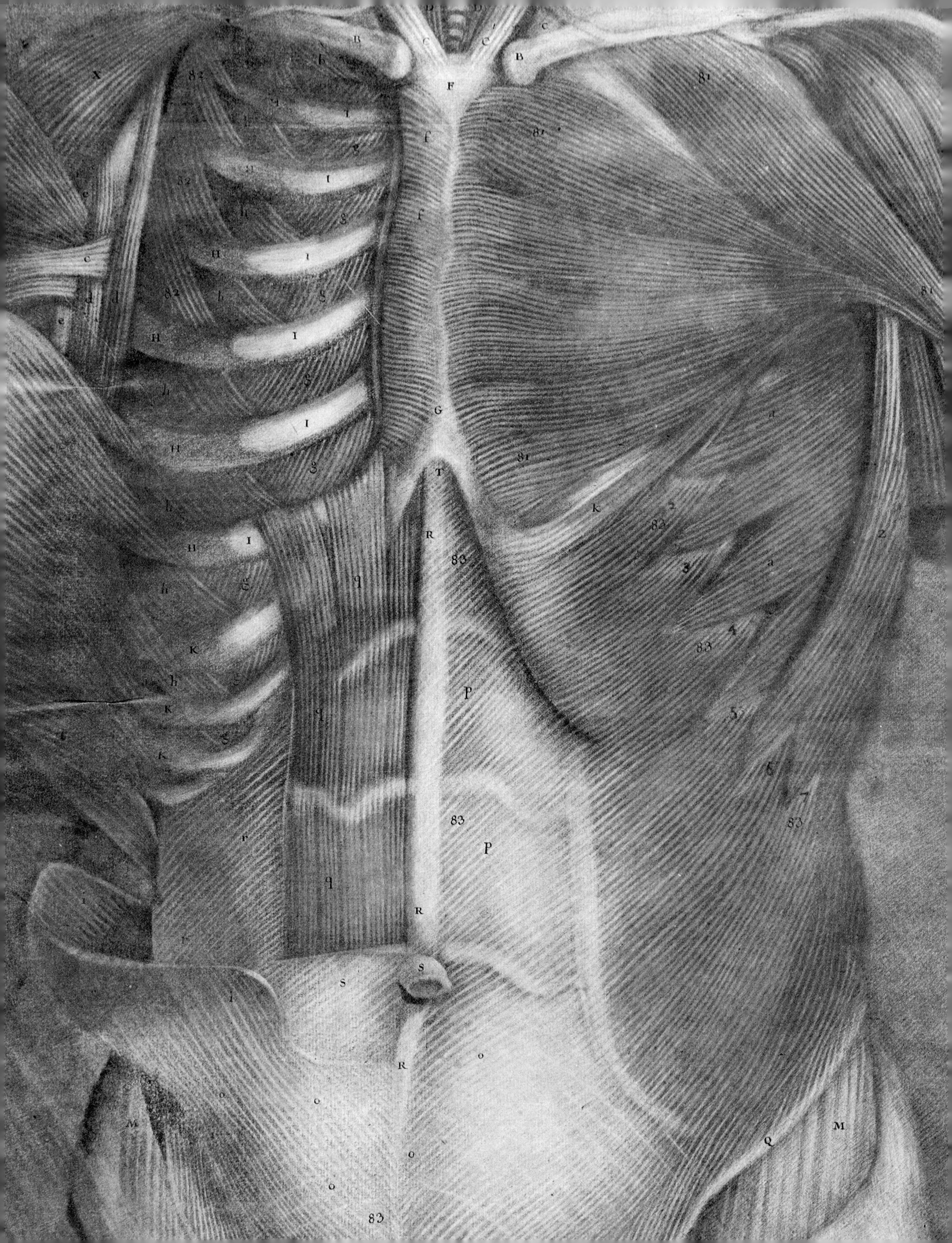

D
D
B
C
C
C
B
X
82
81
F
f
81
82
f
H
82
H
81
G
H
81
T
K
82
R
83
H
3
Z
q
4
83
K
P
K
5
K
6
7
83
83
P
q
R
S
S
R
M
Q
M
83

Chapter 2

The Muscle Machine

Stripped of its covering, all muscle looks much the same. Its surface, as this eighteenth-century mezzotint reveals, is sinuously uniform. But a mechanical wizardry lurks beneath, at microscopic levels. Hidden from the naked eye, inside muscle's meaty bulk are its secrets of contraction, the source of all human motion.

Muscle moves. And by its motion we move. Yet despite the variety of actions we are capable of performing, muscle itself moves only by contraction. Streamlined, muscle is a single-purpose machine with a task as obvious as the ripples beneath an athlete's skin and just as mysterious as the spark that makes an athlete great.

A muscle cell is not the only form of life gifted with motion, but it is the most gifted mover. An amoeba slides about, using its jellylike extensions as arms to pull its body along. The cells in a developing embryo migrate to parts of the body where they will later specialize. White blood cells, too, move with apparent purpose as they stalk invading bacteria. And the long, whiplike tail of a sperm cell lashes to propel it. Scientists have discovered that these and many other forms of life have something in common with muscle. They share two special proteins, the same basic ingredients of motion. From cell to cell, even from species to species, these proteins bear a striking similarity in structure and function.

As a machine for motion, however, muscle is superbly efficient. It normally uses about 35 to 50 percent of its potential energy. By comparison, the automobile engine is a feeble competitor, using about 8 percent. Still, muscle is more than brute machine. Scientists study its mechanism at many levels from the visible to the invisible and, at each, there exists the same striving to contract. The purity of muscle's purpose and design lends it an elegance not usually associated with force. For this reason, muscle holds great appeal for scientists. As a puzzle whose many pieces form complete pictures in themselves, it is an ideal subject for study. Indeed, muscle is so well suited to scientific scrutiny that scientists are using it as a gateway to learning secrets even greater than its movement. As all biologists do, Hungarian scientist Albert Szent-Györgyi, who traced muscle's

The noble heart of ancient science and modern metaphor bubbles not only with blood but emotion. Easily the most important muscle in the body, cardiac muscle tissue powers the heart alone.

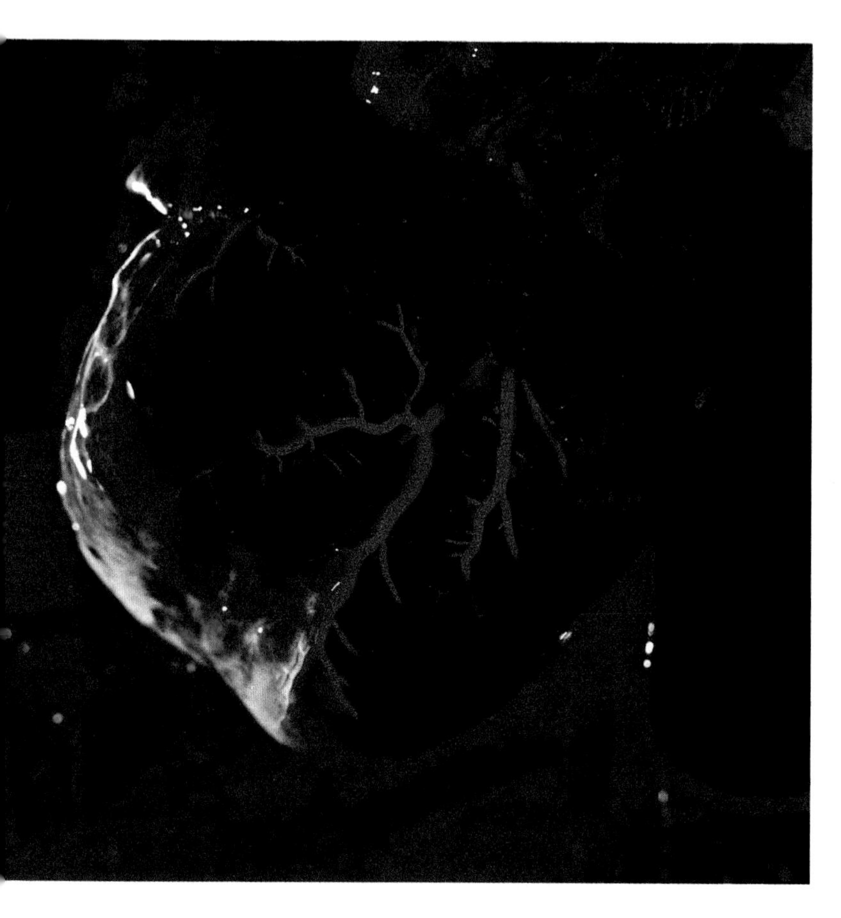

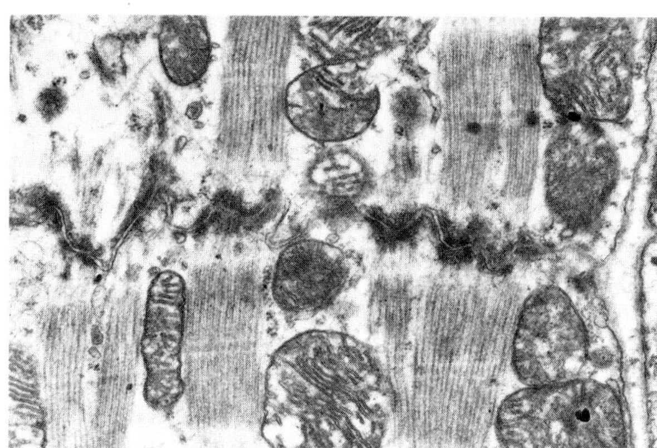

Junctions called intercalated disks, the dark, horizontal markings above, link cardiac muscle fibers end to end. They allow the heart to act as a unified whole by speeding electrical impulses from cell to cell.

contraction to its two proteins, ultimately sought to understand life. Of all Nature's riches, he chose muscle as his mentor.

Special Muscle, Special Task

Although all muscle does perform a single action, the body has different chores that muscle must accomplish and, accordingly, there are three distinct types. Cardiac muscle, found only in the heart, fires lifelong pumping. Smooth muscle surrounds the internal organs and the blood vessels that fuel them. Both are called involuntary, meaning that they usually are not consciously controlled. Skeletal muscle, the third type, carries out our voluntary movements. The different types of muscle govern different worlds. All are specialized, even their cells, or muscle fibers, so named for their slender, cylindrical shape.

Many intricacies of muscle have been revealed only in the last few decades, provided by the technical advances that have enabled man to see ever smaller living structures. Until the invention of powerful electron microscopes in the 1950s, scientists had thought that heart muscle acted like a single cell containing many nuclei, because they could not see the delicate membranes dividing the tissue into cells.

Typical cardiac muscle fiber is short, branching and completely encircled by a membrane. The membranes fuse where the cells meet end to end, forming junctions called intercalated disks. These disks, appearing as horizontal markings across the fibers, provide a strong bond between cells. They also allow the cells to pass electrical signals that tell neighboring cells to contract.

Scientists think that all cardiac fibers might have the potential to contract spontaneously. In some of the heart's muscle fibers, however, this ability is highly developed. The impulse to beat begins in the heart's natural pacemaker, the sinoatrial node, and spreads through a network to reach Purkinje fibers in the heart's ventricles. Branching into the heart wall, the Purkinje fibers carry the pacemaker's transmitted message to ordinary cardiac fibers. And so the heart beats, its cells unified by function.

Cardiac muscle is so harmonious in action that it almost seems to support the abandoned theory

Tiny smooth muscles, piloerectors, tug on shafts of hair to pull them erect, causing the "goose bumps" of a chill. Stained purple, one muscle appears as a fine line next to the hair shaft, at center.

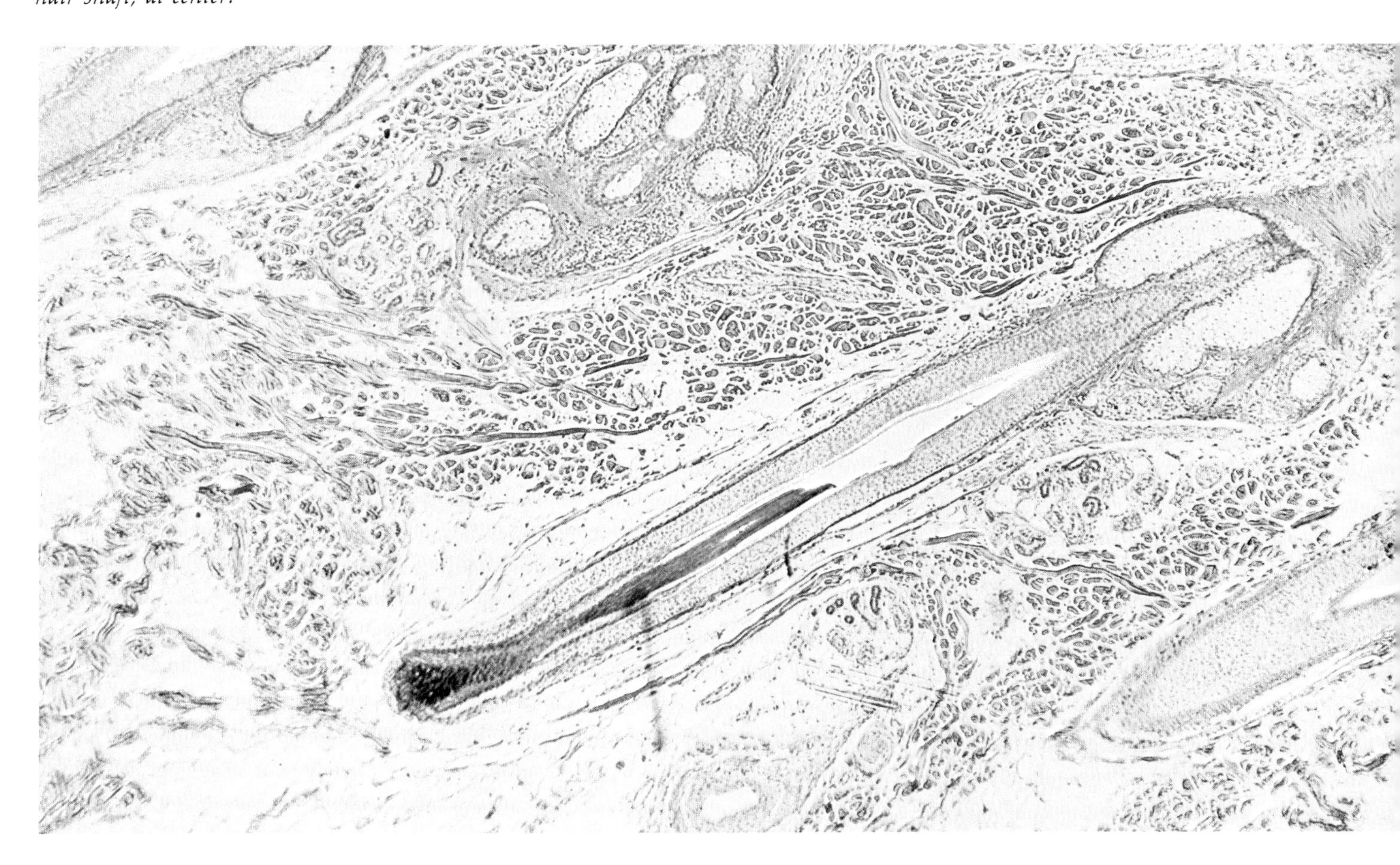

of the single cell. The cells are so tightly bound together and their membranes so permeable to electrical impulse that they act as a mass of merged cells technically known as a syncytium.

Unlike cardiac muscle, smooth muscle has many varied duties. Smooth muscle helps push a baby from the womb, dilates the pupil of the eye and moves food and waste through the body. Its slender, cylindrical fibers are aligned parallel, forming sheets of muscle. In hollow organs like the stomach or bladder, the sheets are usually arranged in two major layers, perpendicular to each other. One forms the outside wall of the organ and the other wraps around the inside. In blood vessels, however, the fibers circle the walls like thread around a spool.

Smooth muscle differs from organ to organ, but there are broadly two types — multiunit and visceral. Multiunit smooth muscle is made of independent fibers separated from each other by a glycoprotein, an insulating substance, and linked by a single nerve ending. Multiunit fibers, controlled almost exclusively by nerves, seldom contract spontaneously. Multiunit smooth muscle encircles many of the large blood vessels. It shapes the eye's lens and makes hair stand on end when we are cold or frightened. A small muscle called a piloerector, attached to the base of each hair follicle, pulls the hair erect upon command from a nerve. This is the same kind of muscle that bristles a cat's fur when it is frightened, making it appear bigger and, presumably, more fierce.

Most smooth muscle is visceral. Surrounding nearly all of the body's organs, its fibers are much more tightly knit than those of multiunit smooth muscle and are connected by junctions known as nexi. The junctions permit them to communicate and, thus, to perform as a single unit much like cardiac muscle.

Like a cobblestone wall, striated muscle fibers, in cross section, are cemented by connective tissue, the endomysium, carrying capillaries to the fibers. The small dark dots are nuclei scattered between the fibers.

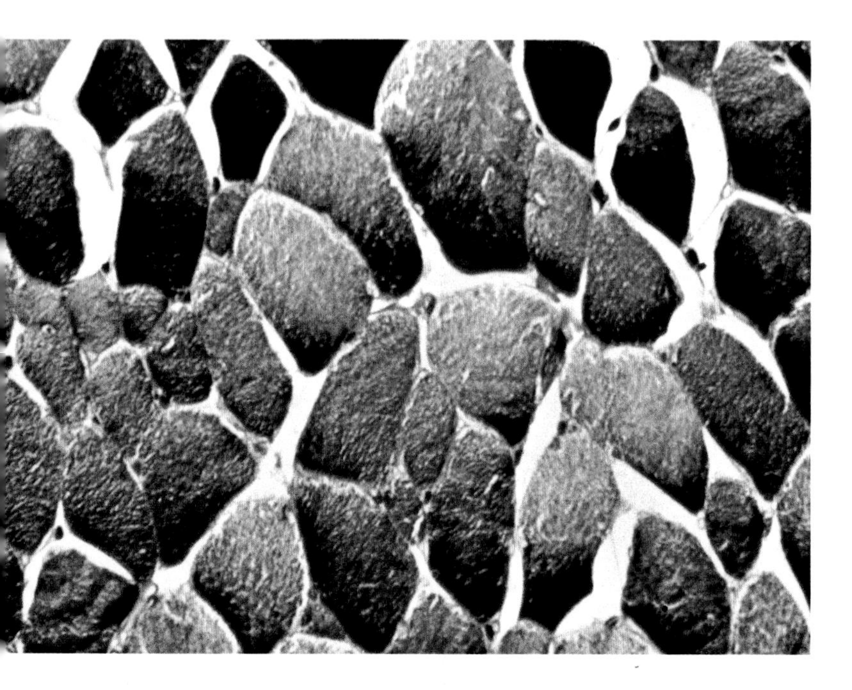

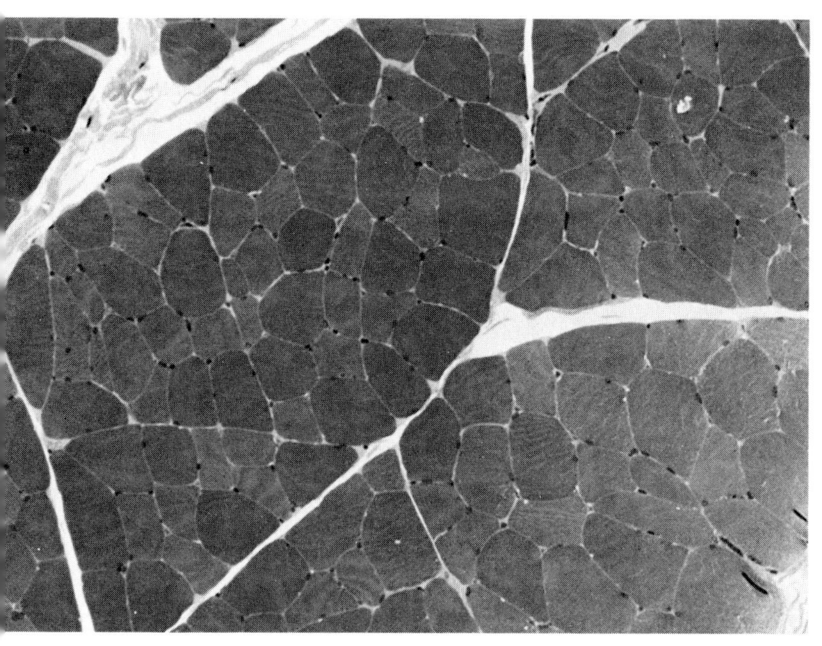

Lower magnification discloses a larger grouping. Connective tissue links the fibers into bundles called fasciculi. In turn, fasciculi are bound together by other layers of connective tissue, the external perimysium.

Smooth muscle possesses two qualities closely related to its work. It has the slowest contraction and relaxation periods of all muscle, and its action is rhythmical. In tubular types of muscle, which is found in the lower esophagus and the intestines, the contraction forms a constrictive ring that moves down the tract, pushing the contents forward. This method of contraction is called peristalsis.

Because many organs encircled by smooth muscle frequently and sometimes dramatically expand, the cells must be able to adapt. Smooth muscles can contract effectively even when stretched twice their normal length or shortened by half. They are also able to maintain a stable degree of tension despite varying degrees of stretch. When a smooth muscle expands, its tension increases but, within minutes, the original tension is regained. Conversely, when the muscle returns to its resting state, it immediately loses tension but regains it within a minute or so. Greater tension can trigger contraction in smooth muscle. Like all muscle, smooth muscle is controlled by nerves and hormones. But only smooth muscle can be activated by stretch, the response to direct mechanical pressure or distortion. Thus, a hollow organ like the bladder, when full, automatically begins to contract.

But the word *muscle* brings something else to mind. What we will into action, what gives the body form, what aches after a ten-mile hike is skeletal muscle. Anchored to bones, it pulls on them to initiate movement. Accounting for 23 percent of body weight in women and 40 in men, skeletal muscle is the body's largest tissue.

Patterned for Power

Skeletal muscle fibers are elongated cylinders containing several nuclei, originally belonging to smaller cells known as myoblasts that merged together before birth. Much larger than other muscle fibers, many skeletal muscle fibers are visible to the naked eye. Some, like those in the thigh's sartorius muscle, are more than a foot in length.

Individual fibers can extend the entire length of the muscle. Usually, however, one end of the fiber attaches to tendon, the tough tissue that binds muscle to bone, while the other end at-

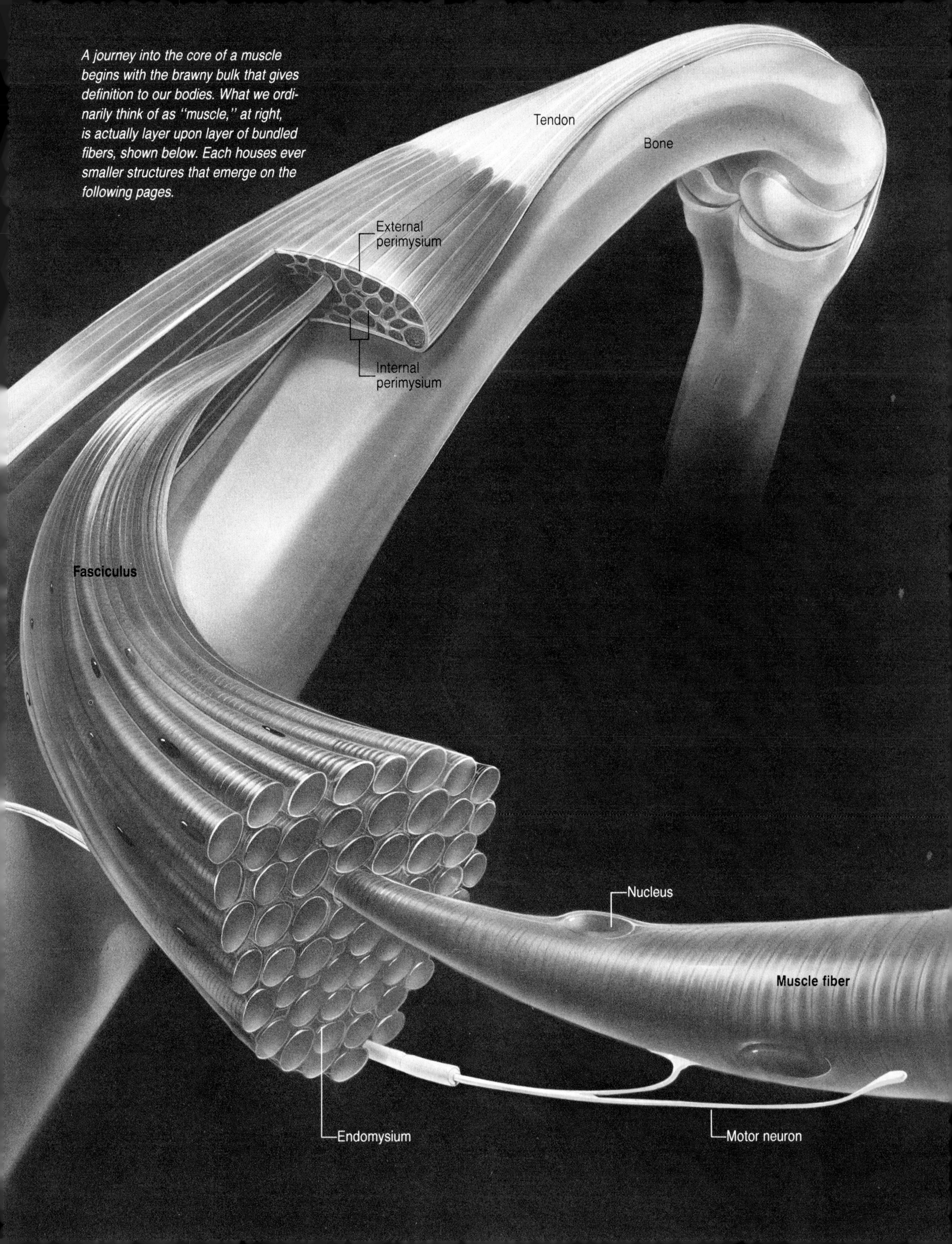

A journey into the core of a muscle begins with the brawny bulk that gives definition to our bodies. What we ordinarily think of as "muscle," at right, is actually layer upon layer of bundled fibers, shown below. Each houses ever smaller structures that emerge on the following pages.

Magnified 100 times, fibers from a skeletal muscle in the abdomen display the precision of their striations. The fine, wavy lines creating a pattern arise from the fibers' smaller components, the many myofibrils.

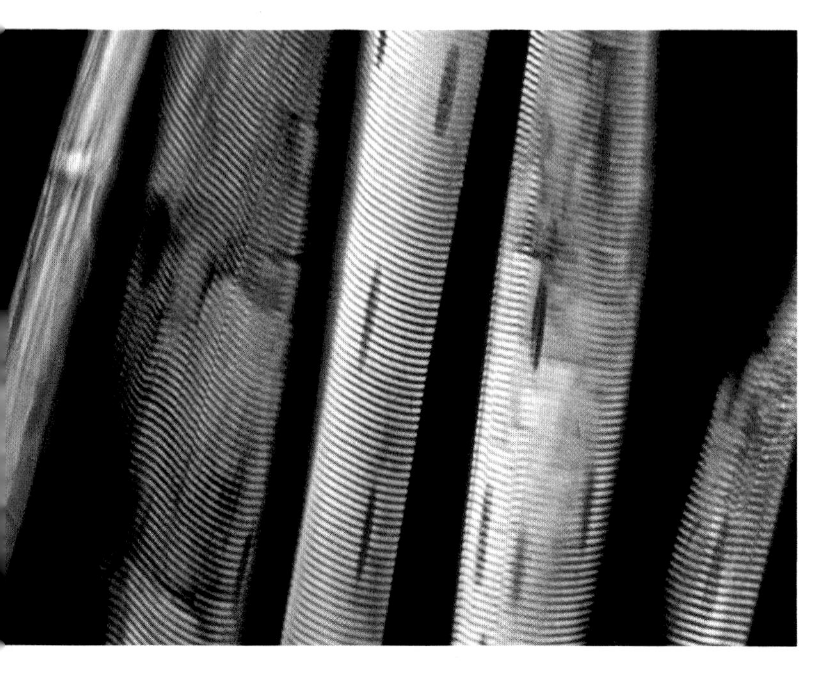

Hundreds, sometimes thousands, of myofibrils fill a single muscle fiber. Thin membranes partition each myofibril into many smaller units known as sarcomeres, the individual units of contraction.

taches to connective tissue in the muscle. In a sense, the firm, white tendons form a core for muscle by extending far inside its mass and emerging at the ends of the muscle to link it to bone. Muscle fibers and tendon fibers are completely different materials, however, and do not merge. Instead, connective tissue extending from the tendon forms a cuplike receptacle for the muscle fiber ending. Other fibers tangle around the muscle fiber, binding it to the receptacle.

Surrounding the muscle fiber is the endomysium, a thin sheath of connective tissue. Another sheath, the internal perimysium, bundles the individual fibers into groups of about twelve called fasciculi. These bundles are themselves bound together by another layer of connective tissue called the external perimysium or epimysium. It is this final grouping that we commonly speak of when using the term "muscle."

The dynamics of muscle, however, are locked in its basic component, the cell. An individual fiber is surrounded by a thin plasma membrane called the sarcolemma. About 80 percent of the fiber is filled with tiny fibrils, or myofibrils — from several hundred to several thousand, depending on the width of the muscle cell. The remainder of the fiber is occupied by sarcoplasm, a

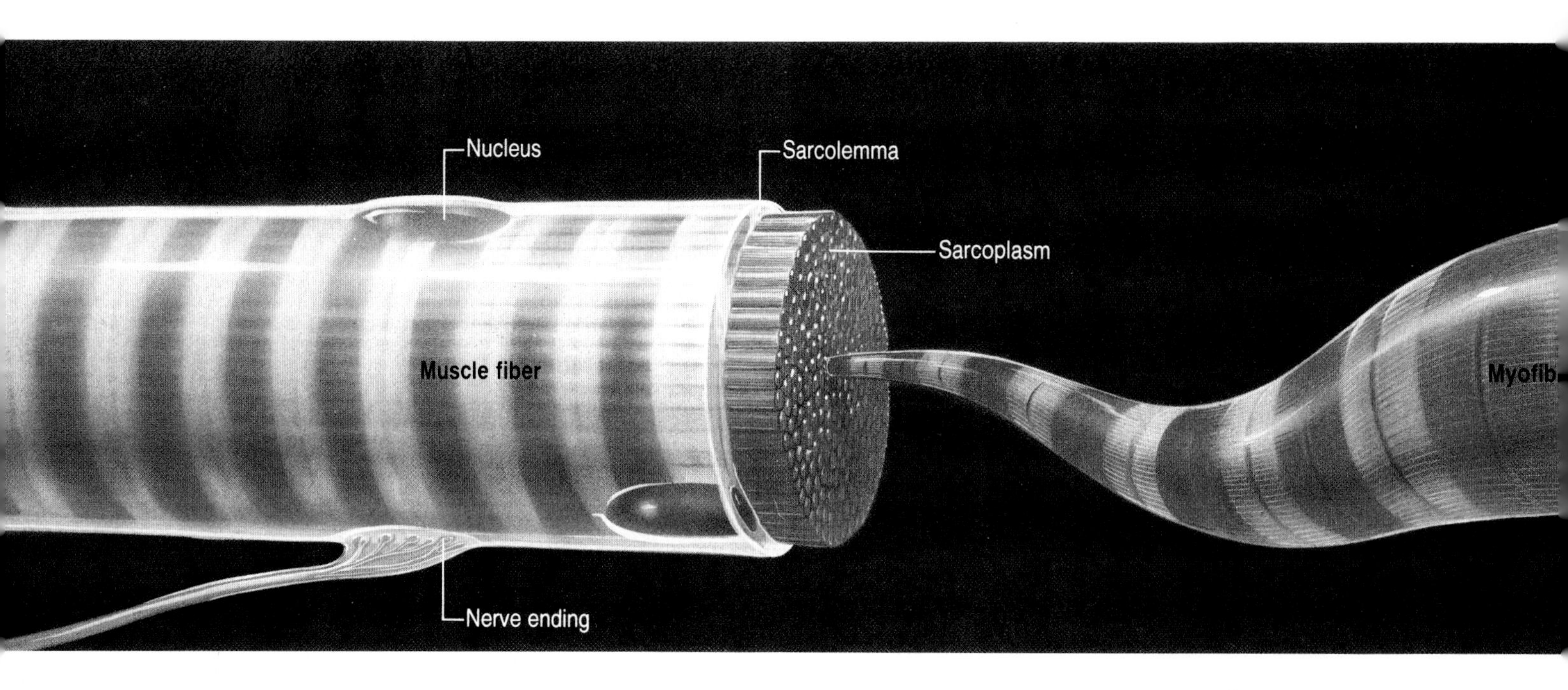

jellylike intracellular substance, the nuclei and other constituents of a typical body cell.

Skeletal muscle is referred to as striated for the simple reason that it is striped. The striations arise not from the surface of the cell but from its many myofibrils, each of which is striated. The parallel arrangement of myofibrils within the cell gives it its characteristic appearance.

First named in German, the markings are known today by letters only. Thin bands called the Z membranes (from *Zwischenscheidung,* meaning "dividing band") separate the myofibrils into compartments. These membranes appear on the fibrils as dark lines. The compartments are sarcomeres, the basic units of a muscle's contraction. Each sarcomere has an identical set of striations. The thin Z membranes form the borders of each cylindrical sarcomere. Next to each Z membrane are light regions called the I bands. A prominent, dark stripe, the A band lies at the center of the sarcomere. The A band itself is striated with a light center, the H zone, and a fine, dark stripe that cuts through the H zone — the M line. This exact pattern is repeated, sarcomere by sarcomere, down the length of the fibril.

These tersely named bands are not just abstract markings. Each signifies a portion of the sarcomere's contractile mechanism, another intricate level of the muscle machine. Just as a single fiber contains many myofibrils, each myofibril contains many smaller filaments arranged in a repeating pattern along the length of the fibril. There are thick filaments composed of the protein myosin and thin filaments composed of the protein actin. The arrangement of the filaments gives rise to the striations. The thick filaments form the dark regions of the sarcomere; the thin filaments form the light areas. Thus, the dark A band in the center of the sarcomere consists mostly of myosin filaments. They are anchored at the M line, the center of the A band. The thin actin filaments, anchored to the Z membranes, form the light I bands. The darkest striations occur where myosin and actin filaments overlap. The A band is darkest because it normally includes both types of filaments. The actin filaments do not quite stretch into the center of the sarcomere, however, which is why the H zone appears lighter.

The filaments, like the resulting striations, make little sense unless connected with their mechanism. A seventeenth-century Dutchman, amateur scientist Anton van Leeuwenhoek, was the first to observe a muscle fiber's striations un-

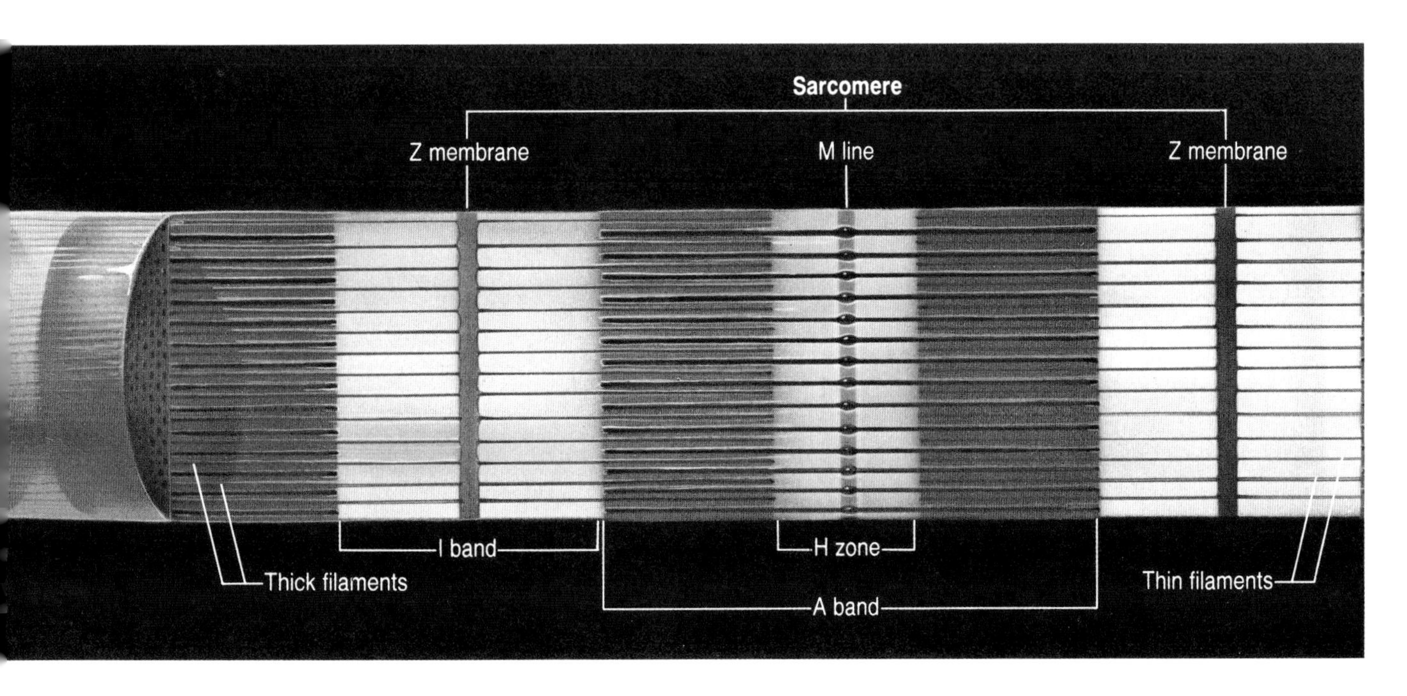

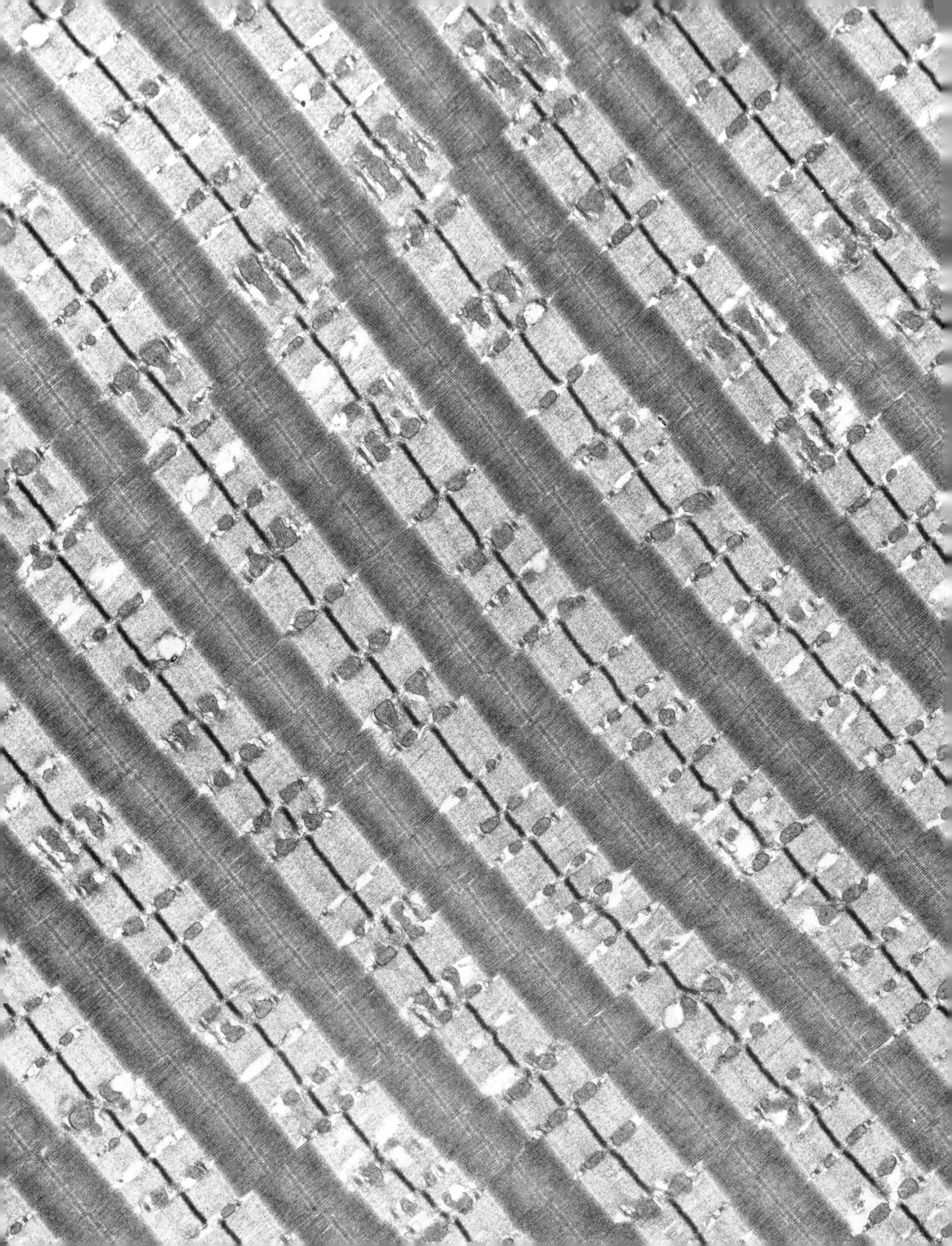

Magnified 12,000 times, human skeletal muscle rivals modern art. The alignment of sarcomere after sarcomere produces a striking series of striations that gives it an air of relentless energy.

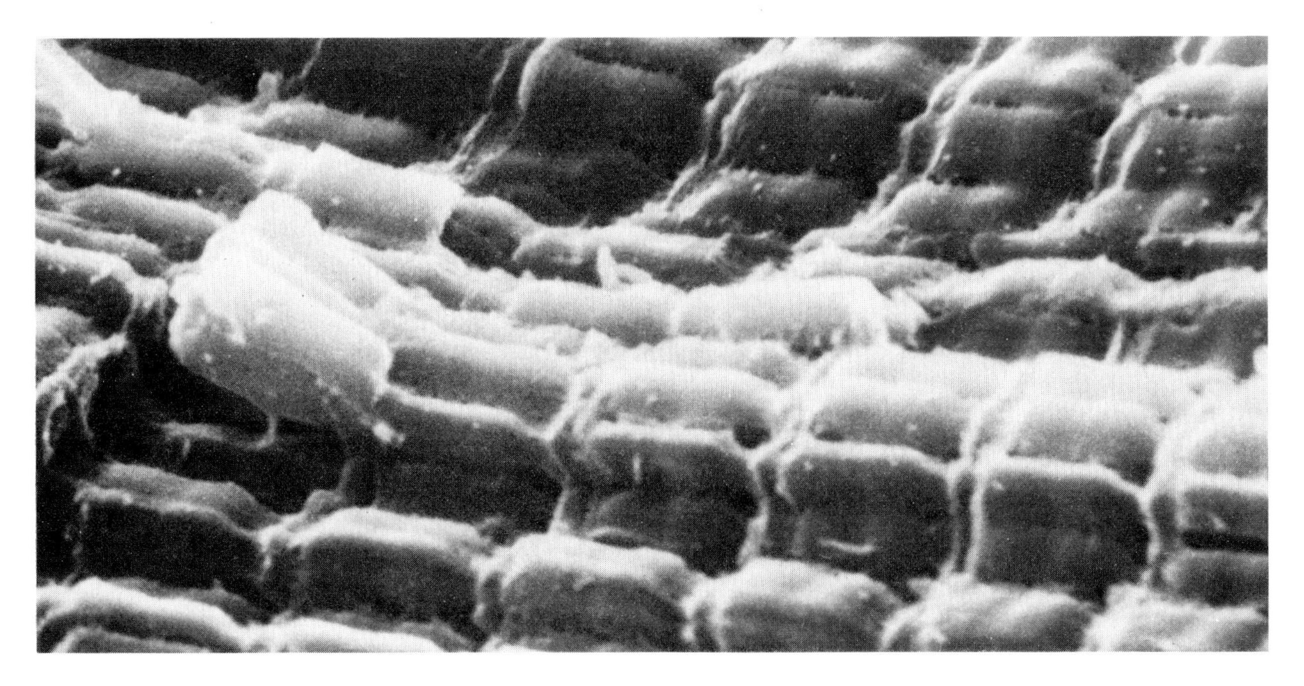

Magnified 26,000 times, the sarcomeres' abstract quality disappears. The broad, dark A bands become smooth, raised cylinders aligned in rows. Crossing them are the thin, light I bands and Z membranes.

der a microscope. He noted that each fiber was "full of rings and wrinkles," which he instinctively felt were connected with the mechanism of contraction. He sensed that the secret was hidden in the smallest reaches of the fiber, smaller even than the tiny filaments he had seen. "Who can tell," he wondered, "whether each of these filaments may not be enclosed in its proper membrane and contain within it an incredible number of still smaller filaments?"

His question would sleep through nearly three centuries. Until the mid-twentieth century, most scientists had assumed that when a muscle shortened, its components did, too. The filaments, they had reasoned, probably folded or coiled during contraction. But when the new electron microscopes provided a glimpse of muscle never seen before, researchers knew that they had to look for another mechanism of contraction. They found that when a fiber contracted, the length of its dark A bands remained constant. Since the length of the A band is equal to the length of the thick filaments, scientists realized the thick filaments did not change in size during contraction. The two light regions, the I band and the H zone, did shorten when the fiber contracted. Evidence suggested the thin filaments corresponding with

Hugh Huxley

Sir Andrew Huxley

Finding the Mechanism of Contraction

The story read like an entry in *Ripley's Believe It or Not.* Two men, working independently of each other in the 1950s, came up with the same revolutionary concept at the same time and, although unrelated, they shared the same surname. Both scientists worked in towns named Cambridge — molecular biologist Hugh Huxley in Massachusetts and physiologist Andrew Huxley in England.

Their theory held that when a muscle contracted, the microscopic components of its individual fibers slid past each other, but did not actually contract. This idea, which soon became known as the sliding filament theory, dramatically changed the way scientists looked at muscle.

Scientists had known that muscle fibrils were striped with repeating light and dark bands, and that the fibrils contained two proteins, myosin and actin. They also had thought the proteins were uniformly distributed along the fibril. In 1953, Hugh Huxley and colleague Jean Hanson found that the protein myosin appeared only in the broad, dark striations, while actin occupied the light bands. This suggested the existence of a mechanism of contraction linked to the fiber's striations,

an idea most scientists had discarded earlier in the century.

The next discovery surfaced on both sides of the Atlantic. It was the most important clue leading to the contractile mechanism of muscle. The length of the dark bands, called A bands, did not change when a muscle contracted, which indicated that the myosin structures did not shorten. In the May 22, 1954 issue of England's prestigious *Nature* magazine, the two individual studies appeared, back to back, reporting this information and proposing a similar theory for muscle's contraction.

Although Andrew Huxley declared the theory "radically new," he also acknowledged that earlier scientists, almost a century before, had collected the same basic clues leading to the theory.

Andrew Huxley believed that history simplified the progression of science. The mere passage of time did not guarantee achievement. His nineteenth-century predecessors had worked without electron microscopes. Lacking the means to confirm their ideas, they relied on light microscopes and their powers of observation and intuition to suggest that myosin filled only the dark A bands and that these bands did not shorten during contraction. But their approach, assuming that structure explained function, had seemed too simple. It became scientifically unfashionable. So their ideas were neglected. Thus, the final coincidence in the Huxleys' theory was one of timing. As important as the technological advances of the modern age were, a change in scientific attitude allowed the theory to flourish.

these regions did not contract, however. Because the distance from the Z membrane (where the light filaments were anchored) to the H zone (where they met in the middle of the sarcomere) did not change during contraction, scientists concluded that neither did the light filaments.

But if the filaments did not shorten when the fiber contracted, how, then, did contraction occur? There was only one possible explanation: the filaments must slide past each other when the fiber contracted, like lines of soldiers marching in opposite directions. This conclusion seemed so obvious that the two leading research teams studying contraction — Hugh Huxley and Jean Hanson of M.I.T., and Sir Andrew Huxley and Rolf Niedergerke of Cambridge University — announced their findings almost simultaneously in 1954. The sliding filament theory, as it came to be known, has served as a basic framework for other researchers, and has been enhanced by their findings. It is still the leading theory of muscle contraction. Although much remains undiscovered, the mechanism, as it is understood, is one of intricacy. Contrary to the images that muscle sometimes inspires, it is an elaborate instrument.

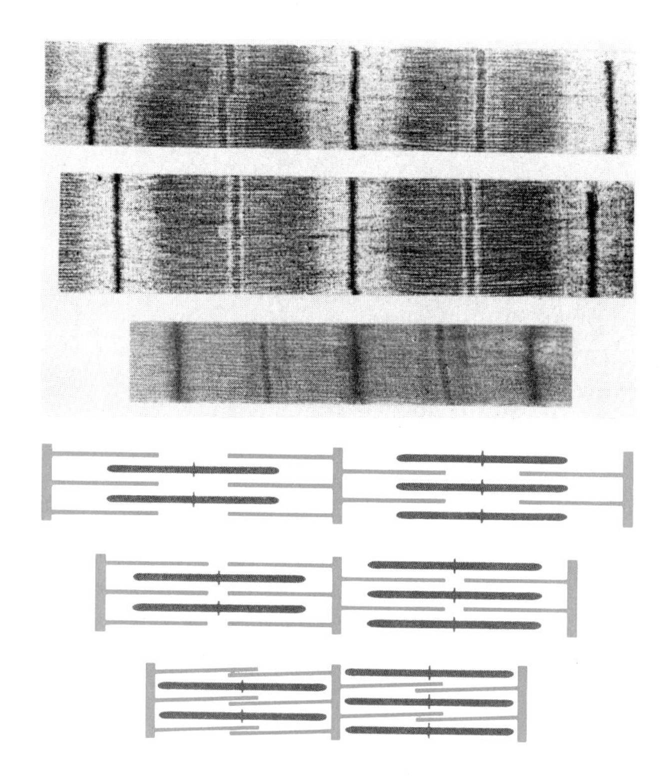

According to the sliding filament theory, when sarcomeres contract their protein filaments, rather than shortening, slide past each other. The changing width of the I bands indicates this movement.

The muscle machine swings into motion when it receives impulses from the central nervous system. The nerves terminate near the cell's delicate membrane, where they release transmitter chemicals. The neurotransmitters initiate a wave of electrical activity that spreads through the whole fiber. This causes the fiber's membrane to release calcium ions, electrically charged calcium atoms that spark the mechanical process of contraction. The calcium ions spread throughout the fiber via a network of fine tubules and diffuse into the myofibrils, coming into contact with the fiber's contractile proteins, the thick and thin filaments. There, they interact with two other proteins, troponin and tropomyosin, which, as a team, circle around the thin actin filaments like delicate embroidery. The calcium chemically binds with the troponin, causing it to somehow influence tropomyosin. The tropomyosin threads shift their hold on the actin filament, uncovering spots along the shaft of the actin filament that are receptive to binding with the myosin filaments. This entire chain reaction transpires in a few milliseconds.

Connecting muscle's thick and thin filaments, cross bridges bring about contraction by repeatedly linking with the thin filaments and pulling them forward. Magnified 220,000 times, resting muscle, at top, looks airy because few cross bridges have formed. With many cross bridges locked in place, bottom, the muscle becomes as rigid as it appears. This specimen shows rigor mortis, the stiffness accompanying death.

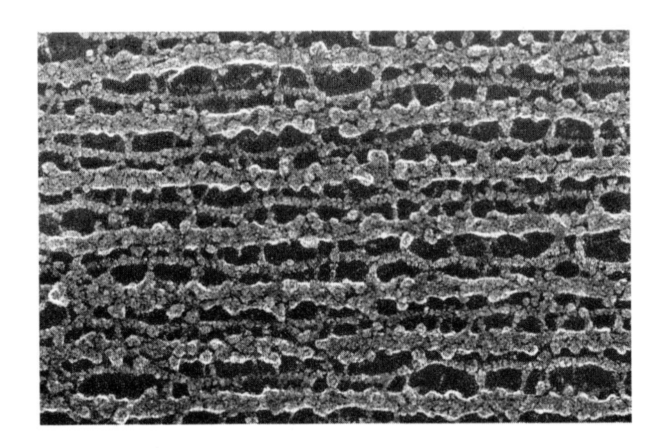

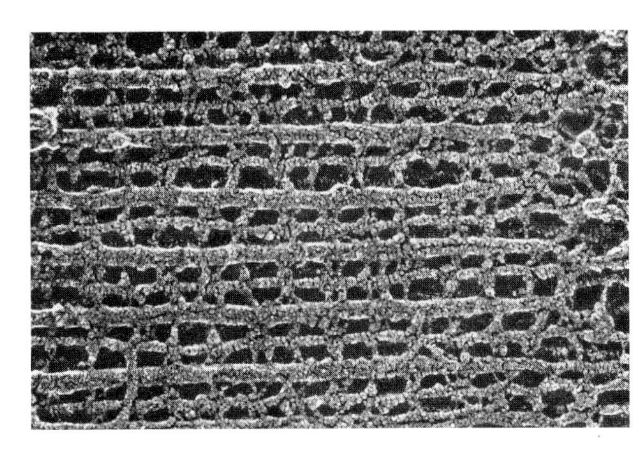

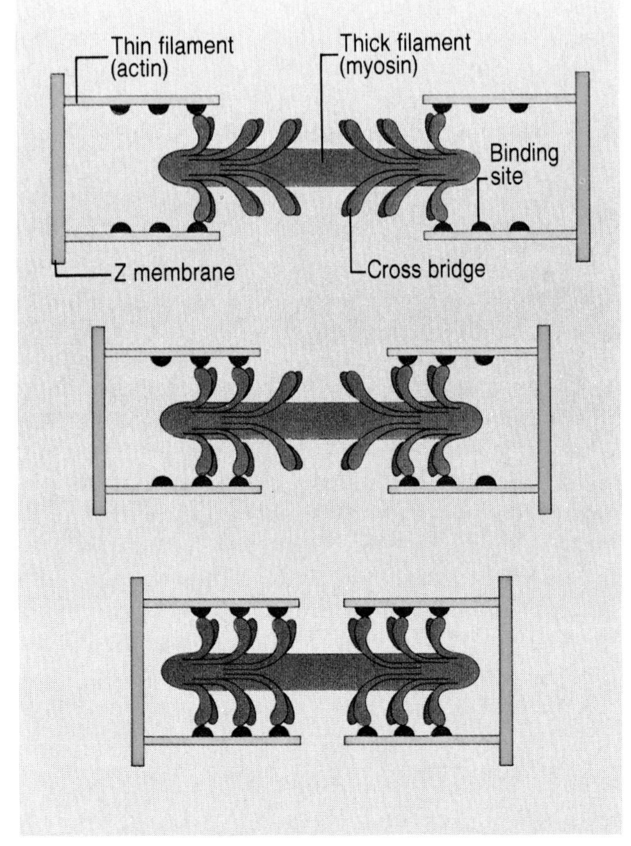

Emerging from the myosin filament are pairs of rounded buds. Each pair forms the head of a single myosin molecule. When he first studied these structures in 1957, Hugh Huxley found them so highly ordered that he wondered if his preparation techniques had "somehow improved on nature." The projections, cross bridges, form the central part of muscle's contractile mechanism and take their name from that role.

The discovery of the cross bridges, the most important addition to the sliding filament theory, was another finding that had awaited advances in electron microscopy and its accompanying techniques. Scientists now think they understand how the cross bridges work in contraction, but the technology to analyze the rapid movement of these tiny structures has not yet been developed.

Cross bridges are crowned with a remarkable substance, adenosine triphosphate (ATP). ATP is an organic compound, the main source of life's energy, derived from food. So vital is this substance that some scientists believe the appearance of ATP on earth may have been the event that led to the appearance of life. ATP provides energy through the loss of one phosphate molecule. The high-energy ATP is thus transformed into two low-energy products: adenosine diphosphate (ADP) — which has two phosphates instead of ATP's three — and the inorganic phosphate that split away. The energy lost in the split is then available for use in the body's metabolism.

ATP and the myosin molecules are said to have a high affinity for each other — so much so that in normal muscle almost every myosin head has ATP resting on it. Each of the head's two buds has a different function. One is adenosine triphosphatase (ATPase), meaning that it has the ability to split ATP and liberate the energy. The other portion of the head binds to the actin filaments, forming the cross bridge.

Radiant sunbursts in crystal form, adenosine triphosphate, ATP, is the chemical compound that empowers the body. The main fuel of all life, ATP puts the cross bridges into motion, allowing muscle to contract.

As the protein tropomyosin shifts away from the binding sites on the actin filaments, the myosin arms are able to link with actin. The myosin ATPase then splits ATP and its energy fuels contraction of the muscle machinery by throwing the cross bridges into action. Without ATP to move the cross bridges, the actin and myosin filaments would remain locked together, the muscle unable to move. This is what causes rigor mortis, or death rigor, the stiffening of muscles shortly after death. Dead muscle cells have no ATP.

But how do the minuscule cross bridges accomplish their work? Since contractions normally shorten a sarcomere by 20 percent or more, it seemed unlikely to scientists that the cross bridges could remain attached to one point on the actin filament. Instead, Hugh Huxley and his colleagues suggested that cross bridges must make and break links with the actin filament in repeated cycles. Like a team working hand over hand to pull a rope, cross bridges hook up with actin, pull the filament forward, then release it and reattach further along its shaft. The cross bridges repeat this action until the contraction is completed. Scientists assumed that one molecule of ATP would have to be split to fuel each cycle. To test the cross-bridge theory, Hugh Huxley worked with a specific muscle in a rabbit's back. In order to activate the muscle, he calculated that the cross bridges would have to go through 50 to 100 cycles each second. This was compatible with the rate at which the muscle used ATP.

The more cross bridges there are the more powerful the fiber's contraction. When a muscle is stretched too much there is little or no overlap between the thick and thin filaments, so cross bridges cannot make enough connections to create tension. On the other hand, with too much overlap, thin filaments on either side of the sarcomere will begin to overlap themselves, inter-

Russian athlete Yuri Stepanov clears a jump seven feet, one inch high for a world's record in 1957. A special shoe aided his soaring performance by slightly stretching a muscle in his takeoff leg.

fering with the action of the cross bridges. And, in a greatly shortened fiber, the thick filaments will be compressed between the two Z membranes, making further shortening difficult. Therefore, a muscle should have one length at which it can contract most efficiently. Normally, resting muscle is quite near this length. The greatest tension is produced when a muscle is stretched slightly beyond its resting length, at which point it will contract more forcefully.

Muscle at Work

Greater contraction can make a measurable difference in athletic performance. In 1957, Soviet high jumper Yuri Stepanov sailed over a high jump for a world's record of seven feet, one inch. Surprise turned to suspicion when photos showed that there was something strange about the shoe Stepanov wore on his takeoff foot. The front portion of the sole was one inch thick and spiked, which effectively lowered the heel. By slightly stretching the large muscle in his calf, the gastrocnemius, the shoe gave him greater thrust at takeoff. So subtle a difference in contraction had boosted his performance. Officials later prohibited its use in competition.

The term "contraction" does not necessarily refer to the shortening of a muscle. Technically, it refers only to the development of tension within a muscle. There are two major types of contraction. A contraction in which the muscle develops tension but does not shorten is isometric, and one in which the muscle shortens but retains constant tension is said to be isotonic. Both are determined by the amount of resistance the muscle meets in contraction.

Someone trying to lift too many books in a box will strain against the weight. His arm muscles develop tension but do not shorten because the amount of resistance offered by the box is greater than the muscles' tension. But when he lightens the load, the working muscles shorten as they contract. This is an isotonic contraction. Because his muscles shorten in overcoming the resistance, the isotonic contraction is said to be concentric. If he wants to put the box down on a table, he must gradually extend his arms. To do so, the biceps lengthen, maintaining tension to counter the

A wounded British soldier lies locked in the painful grip of tetanus. An infectious disease, tetanus produces paralysis by causing sustained contraction. Still difficult to treat, it can be prevented by immunization.

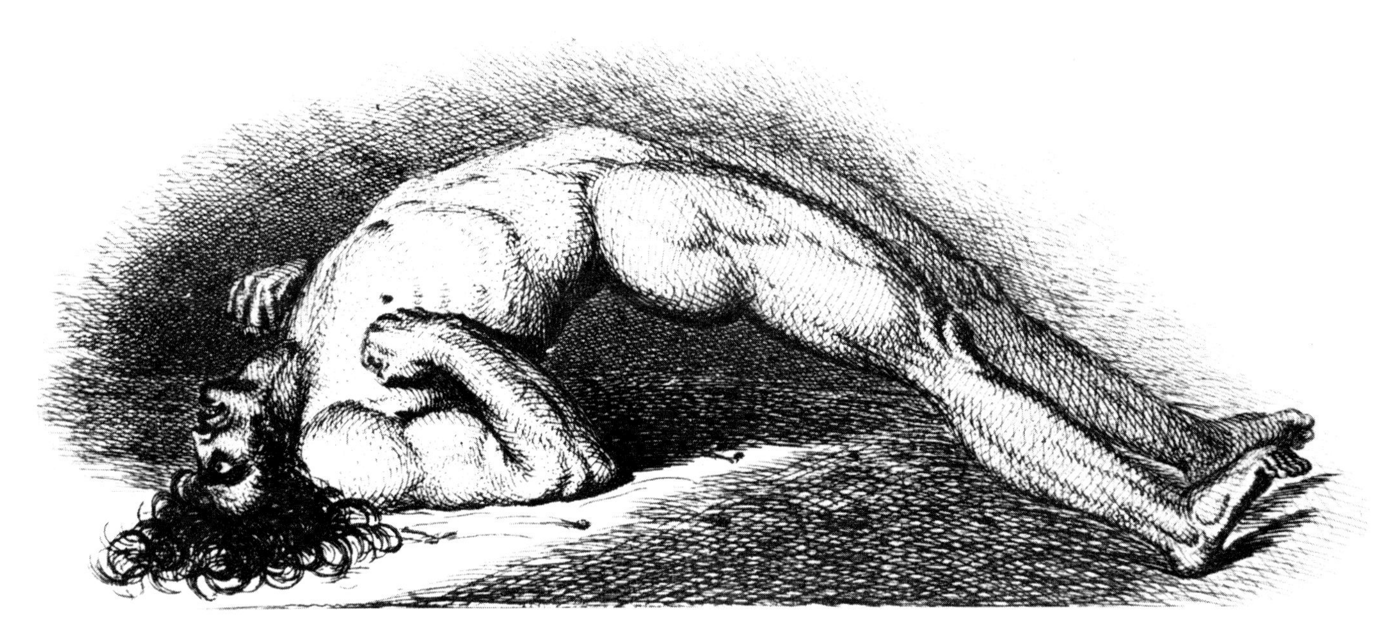

weight of the box. The biceps approach, but do not reach, their resting state. In this case, the isotonic contraction is eccentric because the muscles, in maintaining tension, lengthen.

Muscles must counteract gravity. Although they relax fully, they also remain alert in case they must be called into play to prevent the body from falling. Scientists use the word *tone* to describe this constant state of readiness, determined by a healthy nervous system. Without tone, jaws would hang open and muscles could not support other parts of the body.

When a single nerve impulse strikes a muscle, the muscle's response is a simple contraction, more commonly called a twitch. Each muscle twitch has three distinct phases. The first is a latent period in which the cell starts its chemical processes and the cross bridges begin to develop force. Actual shortening, however, does not begin until muscle surpasses the resistance of its load. The heavier the load the longer the latency. If the load is so heavy that the muscle cannot lift it, the contraction becomes isometric instead of isotonic. The second phase is the contraction. In the final stage, the relaxation period, the muscle returns to its resting state. All three stages take only a fraction of a second to complete.

A succession of individual muscle twitches will produce greater and greater degrees of contraction. This is known as treppe, or "the staircase effect." The contractions will increase in force until they reach a peak, which the muscle briefly maintains, and will then decline. The force of contraction drops with the depletion of the cell's concentration of ATP. The resulting loss of tension is muscle fatigue. Eventually, a stimulus, even a strong one, will bring no response.

If the stimuli arrive in such rapid succession that the muscle has no chance to relax, tetanic contraction occurs. Most skeletal muscle movements, such as the action of the biceps in lifting a box, are tetanic. The rate of impulses necessary to bring about tetanus varies greatly — from 30 to 150 a second — depending on the muscle involved. A normal response of skeletal muscle, tetanus would be deadly if adopted by cardiac muscle, so the heart has built-in measures to prevent it. Cardiac muscle has a longer relaxation period than skeletal muscle and until its muscle is completely relaxed, new stimuli cannot affect it.

The body has no natural protection against another deadly form of contraction commonly called tetanus, an infectious disease unrelated to tetanic contraction. Tetanus bacteria can enter a

Vitality, so much a part of children, slips away when they fall victim to muscular dystrophy, which causes a wasting away of muscles. This child is a participant at a special camp for victims of muscular dystrophy.

minor wound and create a potent poison, one of the most powerful known, which renders a victim prisoner of his own muscles. The toxin acts by blocking nerve impulses that inhibit contraction. Starting at the top of the body and moving downward, the body's muscles lock in paralysis. A rigidly clenched jaw is one of the first signs — hence the term "lockjaw." Still often fatal, tetanus can be easily prevented through immunization and periodic booster shots.

Diseased muscles form an effective prison, prohibiting normal movement and, thus, a normal life. Muscular dystrophy, one leading crippler, accomplishes this by robbing a muscle of its power. Although it is well known, largely due to the Muscular Dystrophy Association's efforts in raising money for research, the disease itself is still quite mysterious to scientists.

Muscular dystrophy is a broad term applied to a group of nine disorders that act in much the same way. Usually hereditary, it has been traced to a flawed gene inherited from either parent. Sometimes, however, the disease can arise from a spontaneous genetic mutation. Muscular dystrophy causes a chronic, progressive degeneration and weakening of muscles. As muscle fibers die, fat and connective tissue take their place.

The most common of the muscular dystrophies is Duchenne, named for Guillaume Duchenne, the French physician who first described it in 1861. The patient, a boy, was seven years old when Duchenne first saw him. The doctor noted with surprise that the muscles of the boy's small legs looked greatly developed, like those of an athlete. They "seemed ready to break through the thin and distended skin." Although firm, even hard to the touch, his muscles "had been deprived of power since birth and hardly had been exercised, the child having a dislike for moving his lower extremities." The false look of

Muscular dystrophy replaces muscle fibers with fat and connective tissue. In contrast to normal muscle, top, the middle photo shows the disease in progression. In its final stages, bottom, most fibers are lost.

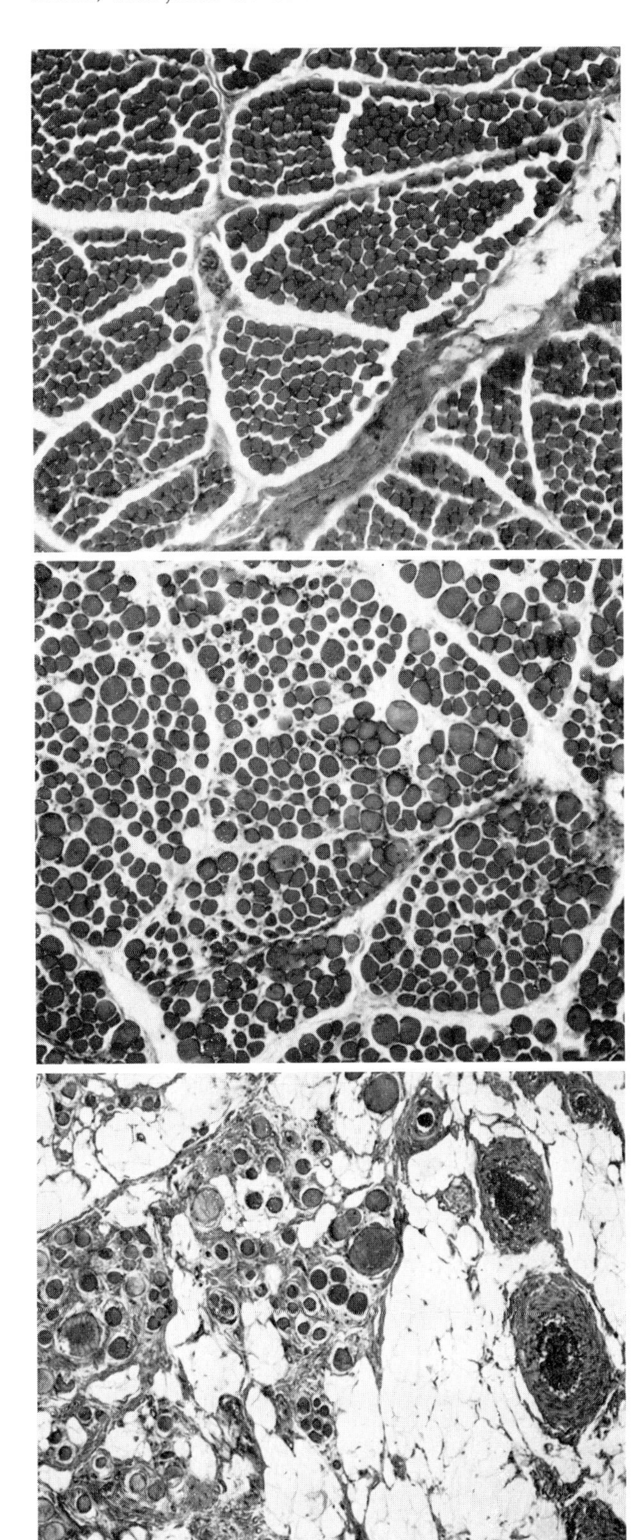

health proved to be evidence of a tragic illness. The apparent growth was caused by a build-up of fat and connective tissue in the muscle.

Duchenne muscular dystrophy is inherited through a flawed gene that is carried by the mother and passed on to her offspring. A female child stands a 50 percent chance of becoming a carrier. A male has a 50 percent chance of inheriting the disease, which strikes about one in every 3,700 boys. It generally becomes apparent between the ages of two and six, when a child's activity and mobility increase. Its first signs are difficulty in standing and running. Usually progressing rapidly, Duchenne cripples victims by the time they reach their teens. Few reach their third decade of life.

Myotonic muscular dystrophy is the most common adult form of the dystrophies. It may occur at any time, but, most frequently it attacks men and women in their twenties. Myotonic dystrophy first affects muscles in the face, neck, hands, forearms and feet. Unlike other dystrophies, muscular deterioration is accompanied by myotonia, a muscular rigidity. The disease progresses slowly, usually not becoming severe until fifteen to twenty years later. Some victims might notice nothing more than a nagging muscular weakness. Often, however, myotonic dystrophy causes some physical disability.

Two other common forms of muscular dystrophy are named for the parts of the body they first affect. Limb-girdle dystrophy first strikes shoulder muscles or muscles in the lower trunk and upper legs. If it affects the shoulders first, the disease progresses more slowly. It usually appears in late childhood or early adolescence and, by middle age, its victims are unable to walk. Facioscapulohumeral dystrophy begins in the face, shoulders and upper arms. The major sign is a weakness in facial muscles. The progression of the disease is often slow, seldom shortening lifespan, but its severity varies greatly. Sometimes its victims are confined to wheelchairs. Physical therapy and orthopedic devices can help keep a patient comfortable and mobile for as long as possible. Despite costly research programs, scientists have not yet found a cause or a cure for muscular dystrophy.

"Muscle man," the colloquial expression for a body builder, has obvious origins. Perhaps the most famous muscle man ever, Arnold Schwarzenegger won thirteen world championships in body building. His greatly enlarged muscles are due to a build-up of contractile proteins in his muscle fibers.

Scientists study muscle enlargement by hanging weights on the wings of chickens. Fibers in the weighted wings are much larger, bottom, than those from the corresponding wing muscle that carries no weight.

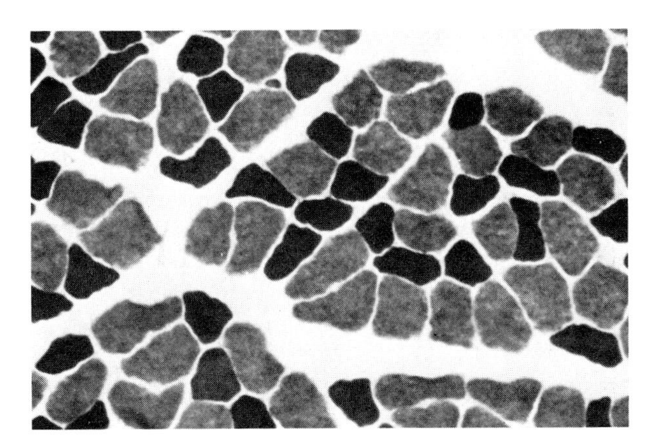

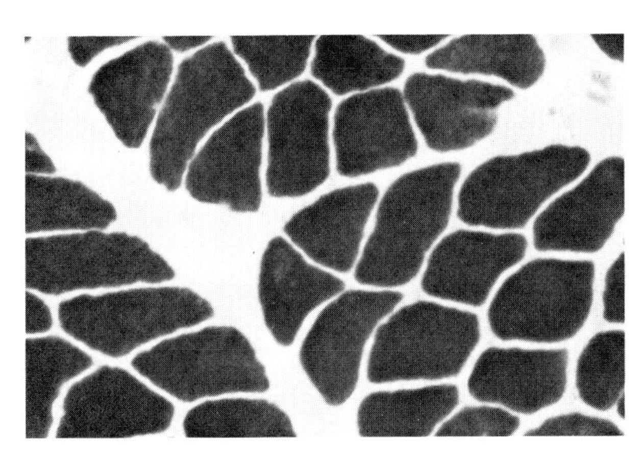

Recent discoveries suggest muscular dystrophy might be linked to an abnormality in the muscle fiber membrane. In the Duchenne variety, the most widely studied, the membrane's repair mechanism seems to be faulty, perhaps allowing important enzymes to leak out. There are abnormally high levels of muscle enzymes in the blood of Duchenne patients. Dystrophic muscle cells also contain other enzymes called proteinases, which break down muscle protein. The Food and Drug Administration recently approved the testing of leupeptin, a chemical that inhibits the action of proteinases. Leupeptin has been shown to delay muscle degeneration in laboratory animals. Researchers also suspect that both myotonic and limb-girdle muscular dystrophy might be linked to a severe deficiency of growth hormone. They hope to find methods of stimulating its production in the body and of supplementing it.

Even a body untouched by muscular disease cannot replace damaged muscle fibers. The number of fibers in any given muscle is fixed at birth. A weightlifter has no more than the proverbial ninety-seven-pound weakling, but he does have larger muscle fibers and muscles with more connective tissue. Exercise, especially weightlifting, stimulates the production of greater amounts of the contractile proteins actin and myosin, making more cross bridges available to carry out more work. With the increase in proteins, the myofibrils thicken and the fibers expand. The enlargement is eventually reflected in the muscles themselves. Primed for heavy work, they are well-defined under the skin. Muscle enlargement, technically called hypertrophy, is not as great in women as in men because it is partially regulated by the male sex hormone, testosterone.

The heart, too, is subject to hypertrophy from exercise. In endurance athletes, the cavity of the left ventricle, the chamber that pumps freshly oxygenated blood back to the body, enlarges. The larger cavity enables the heart to send more blood to the body, powering an athlete's prolonged effort. On the other hand, in athletes trained for spurts of strength rather than sustained effort, the left ventricle's wall thickens. Thus, sprinters and weightlifters cope well with a sudden rise in blood pressure when performing.

Muscle can atrophy, too. When a muscle's nerve fibers are destroyed, the muscle ceases contracting, the actin and myosin content in the fibers fall and the fibers themselves grow smaller. Even if the nerve fibers remain intact, a muscle will atrophy without use. A leg long immobilized in a cast usually suffers some muscle atrophy.

The greatest cause of muscle atrophy is the progression of time. Muscular strength usually peaks at the age of thirty. As people get older, myofibrils degenerate and the number and size of muscle fibers dwindle. Connective tissue replaces the lost fibers, making muscle more rigid and slowing its reactions. The reduction diminishes performance. The process is not as unforgiving as it seems, however. Steady exercise is a valuable measure of preventive medicine that may also help delay fiber loss and maintain strength. By and large, muscle is an exceptional machine, one that works better the more we use it.

Chapter 3

Six Hundred Engines

On July 27, 1976, a mountainous man emerged from his cubicle at the sports arena in Montreal, Canada to survey his competition, the other Olympic weightlifting contestants. He commanded their attention. The herculean body with its barrel chest, arms branching like ironwood and legs like banyan stumps belonged to Vasily Alexeyev, the Soviet weightlifter said to be the strongest man in the world. Nicknamed the Red Bear, he had already won six world titles, an Olympic title, and set nearly eighty world-weightlifting records. He had come to Canada to gain a second Olympic title. His assistants kept track of the time remaining before Alexeyev's first attempt to lift a 507-pound barbell chest high, then heave it overhead, in the first of three lifts allowed in the clean-and-jerk event.

The stress and excitement of this Olympic competition provide a rare glimpse of human musculature pushed to its limits. To prepare his body for the contest of mind and muscle against metal and gravity, the Russian makes an easy lift in the warm-up area. The room rumbles from weights slamming on the concrete floor. With heart and lungs pumping vigorously from the practice, his blood surges with newborn vitality.

Completing the ritual of preparation, Alexeyev returns to a hard wooden table where masseurs pour oil over his massive shoulders and pound the small of his back to warm and loosen his muscles. Signaled that his turn has come, the Red Bear moves his 345-pound body off the table and onto the platform. He is the favorite of the crowd now hushed in heavy silence.

He adjusts his trunks and thick leather belt designed to keep internal organs intact and prevent his spinal cord from snapping beneath the weight of the lift. Dusting his hands with talc and spitting on his palms, Alexeyev glowers at his leaden charge and bends over to test the bar.

Thomas Hart Benton celebrates human potential as steely strength incarnate. Power-driven engines, both mortal and metal cast, labor at backbreaking tasks. A man with heaving muscles — vital engines more than 600 strong — grunts and sweats as pneumatic drills and hoisting cranes clank. Machinery extends man's muscular inheritance, increasing the power that pulsates within his breast.

His tremendous belly, so large that tying his shoelaces is a great effort, intrudes. The royal girth is necessary, though, to provide stability for the lift. Fat, the binding material that keeps his body from falling apart under the terrible load, accounts for 39 percent of his body weight. His tremendous hands, overlaid with blood vessels, nerves and fourteen tendinous muscles that course up to his forearm, engulf the bar. The small intrinsic muscles of his fingers and the short muscles of the thumbs, which normally allow for a wide range of dexterous movements, lock against the knurls that bite into his palms. The grip is perfect. A flat band of six muscles in his forearms works with other muscles about the elbow joints to flex the wrists into position like vises. His huge legs are spread apart, giving him a steely support. As he stoops over the bar, his thigh muscles contract. The adductor magnus muscles in his legs shorten, stabilizing limbs in descent. The muscles about the knees, hips and ankles tighten while others relax, as he lowers himself into squatting position. The lights, the spectators and the judges dissolve as he concentrates on the superhuman feat that he has convinced himself he can perform.

A Champion's Burden

The Red Bear inhales. The musculofibrous diaphragm extends in his thoracic cage, forcing air into his lungs. In an effort born of years of training, his muscles become electric. The diaphragm contracts violently as the musculature of his legs and thighs pulls upward. The iron barbell begins to rise. His hamstrings, long fleshy muscles that course vertically from hip to knee joints, extend as the thigh muscles, interwoven with bands of connective tissue that is both fibrous and tendinous, contract to pull muscle against bone. The knee joints, concerned with weight bearing and preservation of balance, become a battleground between flexing and extending muscles that together ensure a stable, smooth ascension of the trunk. His feet, strapped into leather boots, remain stationary. Any movement would mean disqualification. Strengthened by musculature, the toes wiggling inside the shoes maintain body balance. The long muscles of the legs and short muscles of the feet engage in his effort to lift the seemingly immovable weight. Bones and their points of articulation, the joints, are afire, bearing the weight of the body above and mocking the gravity that obstinately pulls downward.

His knees would extend beyond normal limits except for the strong ligaments and the muscular pull of the hamstring muscles that flex the knees. As he lifts the barbell higher, the transversus abdominis and other muscles that spiral about his trunk, much like the thin laminated sheets in a plywood panel, constrict and compress his internal organs. The broad latissimus dorsi muscles in the lower half of his back move the shoulder joints, transmitting the weight from his trunk to his lower back muscles. A result of the fusion of many groups of small muscles spanning the vertebral segments in the spine, the long, prominent mass of muscle across the midline of his back braces, contracts and stretches.

The biceps brachii, prominent muscles on the front of the upper arms, and the pectoral muscles burst with power as they elevate the arms. The triceps on the underside of the arms stretch, bending the elbows and moving the bar closer and closer to the Red Bear's chest. At the same time, shoulder muscles, the levator scapulae and the rhomboid, counter the pull of the gigantic pectoral muscles in his breast to act against the resistance of the enormous weight.

As the Red Bear lifts the bar to his chest and allows his pectoral muscles to bear the brunt momentarily, he gulps air. With elbows tucked tightly against the walls of his body, arm muscles are freed of some of their weight. Scalene muscles in his neck elevate the first and second ribs and suspend the thoracic cage surrounding his bulging lungs. These accessory respiratory muscles in the abdomen lift the thorax and increase breathing capacity during forced inhalation. In a final huff, nasal muscles dilate in an attempt to admit more air. The air pressure in his lungs quadruples, the arteries constrict and his blood pressure skyrockets. Heart and brain are bereaved of fresh, oxygenated blood. The triceps and the pectoralis pull, slowly extending his arms while the serratus anterior below the armpits rotates his shoulder girdle. Regulating the smooth action,

Vasily Alexeyev, nicknamed the Red Bear, braces himself for an awesome fight of muscle against gravity and steel. With a violent surge of power, he lifts a weight of more than 500 pounds to his chest. Pausing but a moment, he brandishes the barbell above his head in a successful clean-and-jerk.

Two years after pulling a hip tendon, Alexeyev fails at an Olympic comeback. He lifted the barbell shoulder high but fell backward in two attempts. On his last effort, the bar hardly budged off the floor.

Superficial muscles of the chest and arm weave an intricate brocade of living tissues that seldom act independently. A dozen muscles in each shoulder effect the joints' wide range of mobility.

the deltoid muscles stabilize the shoulder joints as his arms lift to a vertical position. Powerful supraspinatus muscles that protect the upper arms from dislocation help pull his arms into complete extension. Leg muscles again push against the platform as the Red Bear straightens. At the same time, the biceps and triceps contract powerfully to extend the elbow joints. Stable, straightened elbows are pushed upward. The saw-toothed serratus anterior muscles on the outer surface of his ribs arch as his arms are stretched to their limit.

His expression becomes a stage on which the dramatic contest between gravity and will, weight and muscle is acted out. The risorius muscles bare the teeth and the corners of the mouth pull sideways in a grimace. The buccinators that extend from molar teeth to the corners of the mouth clench. With lip muscles pulling close against teeth in determination, the platysma muscles surface like cords in his neck.

His body resounds with harmony. Every muscle, every fibril engage, surging with the power of more than 600 engines.

"Lift!" the judge declares, as the audience bursts into cheers. Alexeyev is aware of nothing but the iron bar now bending like vulcanized rubber above his head, his arms locked victoriously into position. The air explodes from his lungs as he pushes the barbell away. The sound of it crashing against the floor is muted compared to the roar that later echoes through the stadium when Alexeyev successfully lifts 562 pounds, clinching the Olympic gold.

This remarkable display of force is only possible through the mortal arrangement of muscles, bones and joints that assemble the body's lever systems. Within the body, bones act as levers and skeletal joints as living fulcrums, points upon which vital powers are exercised. Muscle, attached to bone by tendons and other connective tissue, exerts force by converting chemical energy into tension and contraction. When a muscle contracts, it shortens, in many cases pulling bone like a lever across its hinge. Such a movement depends on many factors, including the influences of other muscles which may be contracting simultaneously.

In very powerful individuals the intense forces produced during maximum exertion can wrench tendons from bone, sometimes even snapping a bone as if it were a toothpick. Such jarring force unseated Alexeyev in 1978 when he tried for an unprecedented ninth world title. As he hoisted 529 pounds of weight to his chest, he felt something tear. Doctors said the superheavyweight champion had pulled a hip tendon. Despite the ensuing years of agonizing physical training and mental conditioning, he was never to return successfully to world class competition.

A Well-Balanced Alliance

When heaving a barbell overhead, the deltoid is the most powerful initiator of force. Since it bears the principal responsibility for this action, it is called a prime mover. Like so many muscles in the human body, its tough, sinewy tissues are named for their characteristic shape, which is tri-

In investigations of actual dumbbell lifting, researchers concluded that a forearm in the pronated, or palms down, position can handle only two-thirds as much weight as a forearm in the supinated, palms up, position.

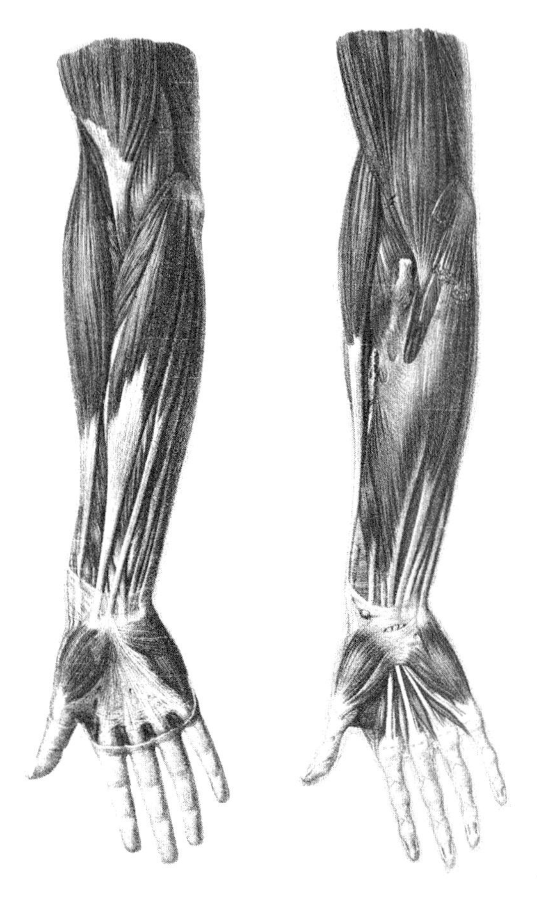

Anterior thigh muscles act with seven other muscles on the knee. The knee joint is perhaps the most vulnerable to injury in contact sports. Most serious ligamentary injuries can be corrected only by surgery.

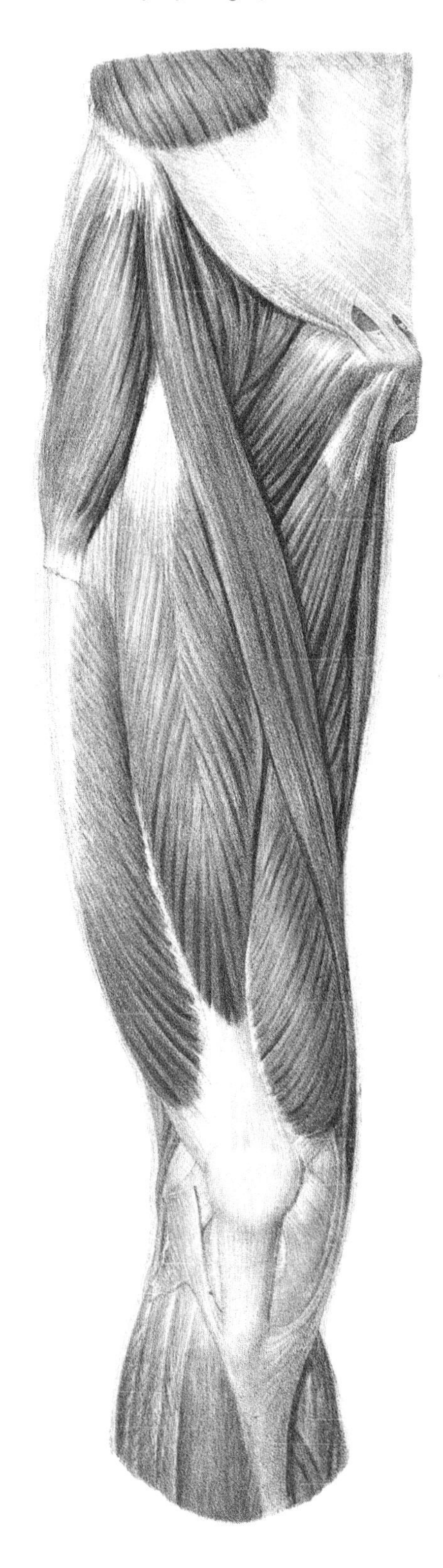

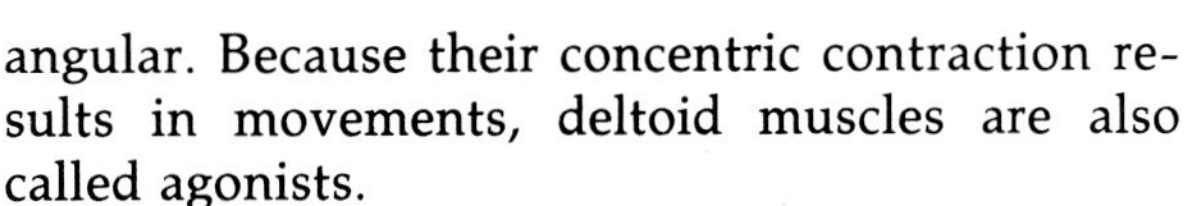

angular. Because their concentric contraction results in movements, deltoid muscles are also called agonists.

An antagonist is a muscle whose contraction tends to produce a joint action opposite to that of the agonist. Any muscle that extends a limb is antagonistic to a muscle that flexes it. The triceps are antagonistic to biceps when the arm is flexed inward, bending at the elbow joint. This arrangement of muscles makes the cooperation and synchronization of muscular effort possible.

Mated muscles faithfully alternate roles as agonist and antagonist in fluid collaboration. When a prime mover contracts, the nerves in its antagonistic opposite slacken tension, stabilizing movement in a well-balanced alliance that scientists call reciprocal innervation. Assistant movers, muscles that contribute to a specific movement, often assist the prime mover in facilitating muscular expression. The hamstring is prime mover

The front of the torso is divided into the thoracic and abdominopelvic regions. Trunk muscles protect the internal organs. To maintain erect posture, abdominal muscles are in a constant state of contraction.

From all directions, muscles converge upon the base of the skull to balance the head above the axial skeleton. Antagonistic muscle action checks head movement largely completed by gravitational forces.

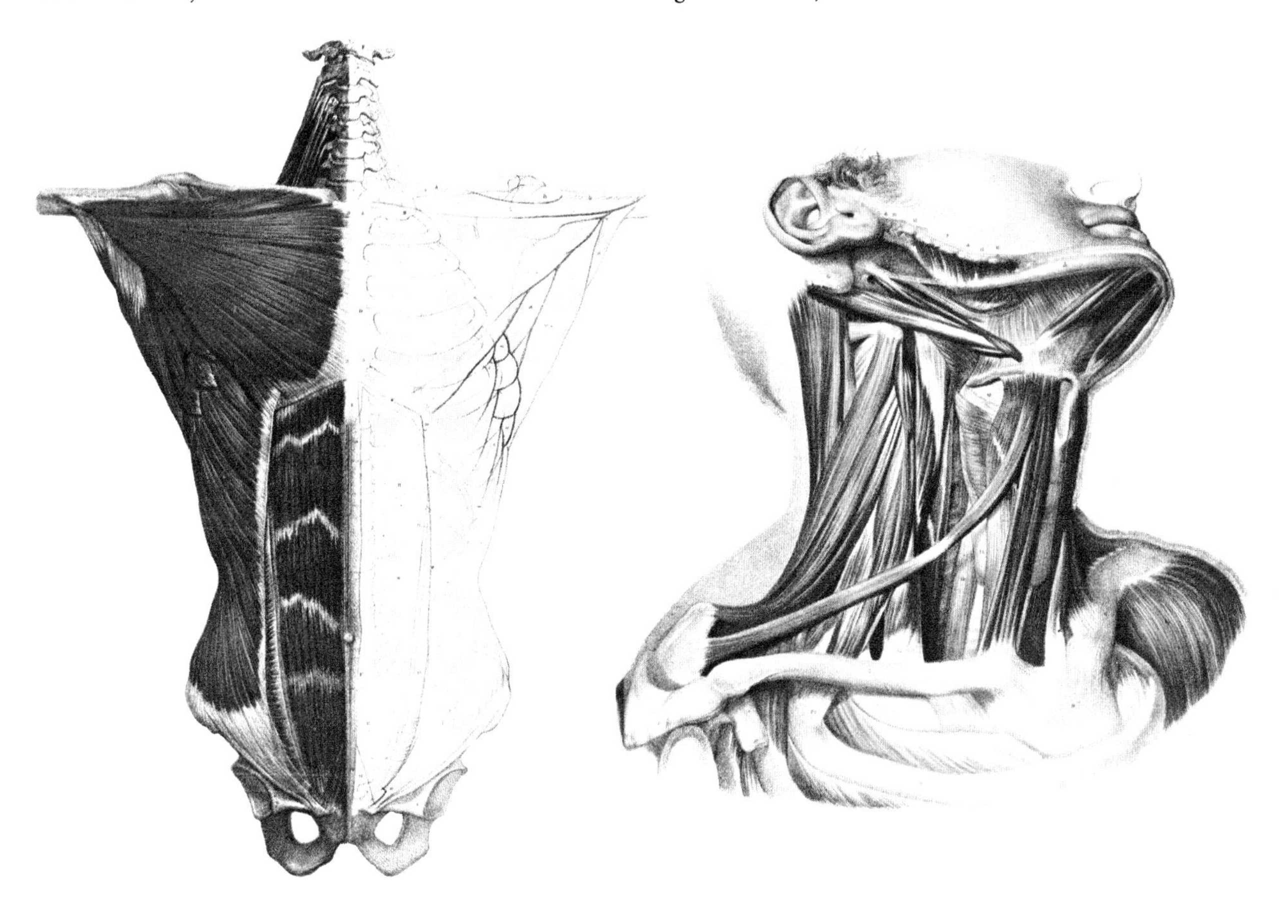

in flexing the knee while the sartorius, the long muscle that permits cross-legged sitting, becomes the assistant and the quadriceps, the antagonist.

Fixator or stabilizer muscles hold a bone or body part steady and so provide a firm foundation upon which active muscles pull. In weightlifting, the abdominal muscles contract statically, preventing sagging of the hip and trunk. This principle known as sequential action successively transfers momentum in coordinated movements.

Synergists act with other muscles to prevent movements that hinder the work of the prime movers. Synergy occurs when one muscle contracts, usually involuntarily, neutralizing other muscles whose domain traverses more than one joint. When a weightlifter grips the bar, wrist extensors act as synergists by holding the wrists extended. If they were to slacken, the long muscles of the fingers would induce wrist and finger flexion, and would stretch the tendons of the

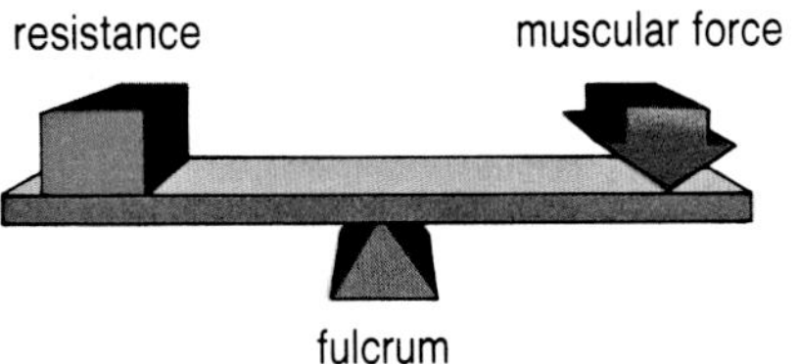

Houston Astros pitcher Nolan Ryan can throw a baseball more than 100 miles an hour. In using his arm as a first-class lever, Ryan's elbow becomes the fulcrum through which force acts upon the resistance.

long finger extensors, causing them to open out and the grip to be lost. Such a situation could have disastrous results in weightlifting.

Extensor and flexor muscles are named for the function they perform. Extensors act primarily to straighten body parts away from the main body, while flexors bend parts toward the trunk.

The extent to which muscle bundles energize depends on the stimulus from the nervous system. Skeletal muscle fibers are cylindrical, multinucleated cells containing contractile threads that shorten when stimulated. For additional muscle force, the nervous system animates more muscle fibers. The impact of these muscle groups working in concert is called power.

Muscles perform work by exerting tension upon the points of their insertion in bones. In all, there are only three types of body lever systems. The zone between the load to be managed and the fulcrum is the load arm of the lever. The region between the fulcrum and the muscle, the power arm, is nearly always shorter than the load arm. Muscles acting on the power arm develop their greatest power by shortening slowly while developing high tension. The force is amplified in the lever system producing rapid movement of the load arm. When the fulcrum is between the force and the resistance, it is called a first-class lever. This kind of lever sacrifices force to gain speed, so that short, relatively slow movements of the muscle produce faster movements of the hand. Thus, Houston Astros pitcher Nolan Ryan can throw a baseball over 100 miles an hour, even though his triceps and forearm muscles shorten at a fraction of this velocity.

In levers of the second class, the resistance is between the fulcrum and the muscular force pitted against the burden. Power rules rather than speed. When a ballet dancer stands on half-point in an arabesque, the entire weight of the body presses against the toes by means of calf muscles. With the toes constituting the fulcrum, the weight of the body acts through the ankle joint and the powerful calf muscles pull on the heel.

In third-class levers, the force is applied between the fulcrum and resistance. The human body comprises a system of third-class levers, machines dependent upon speed for the genera-

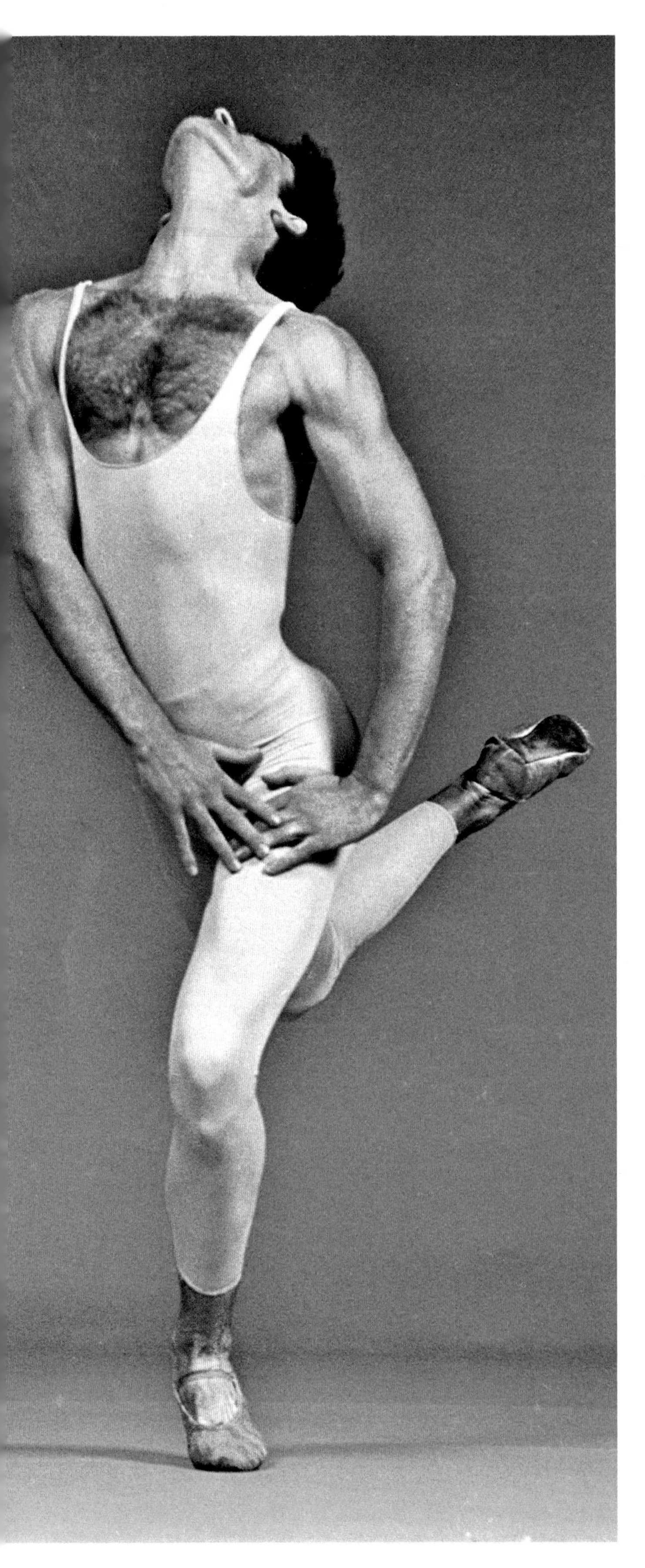

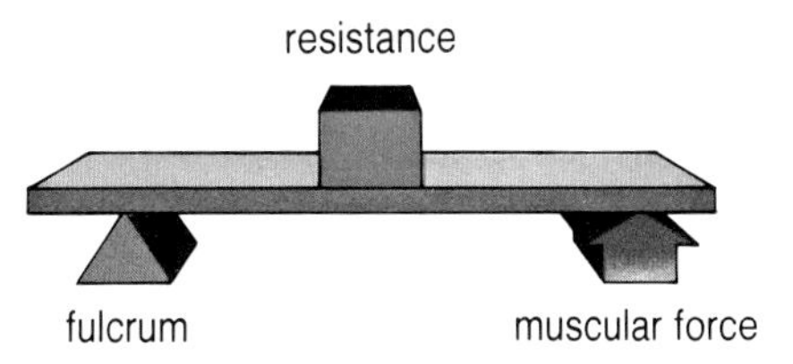

Gregory Huffman of The Joffrey Ballet company demonstrates the body's second-class lever system. To balance in relevé, his toes act as the fulcrum, his weight as the resistance and his calf muscles as the force.

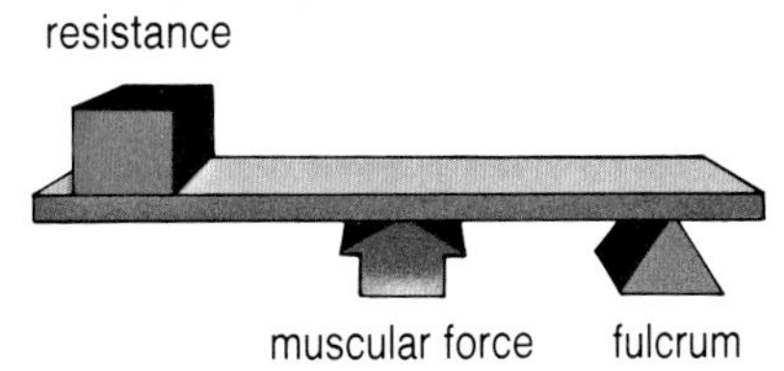

Body builder Franco Columbo performs the biceps curl exercise. Using his arm muscle this way displays the third-class system of body levers in which the force is exerted between the resistance and fulcrum.

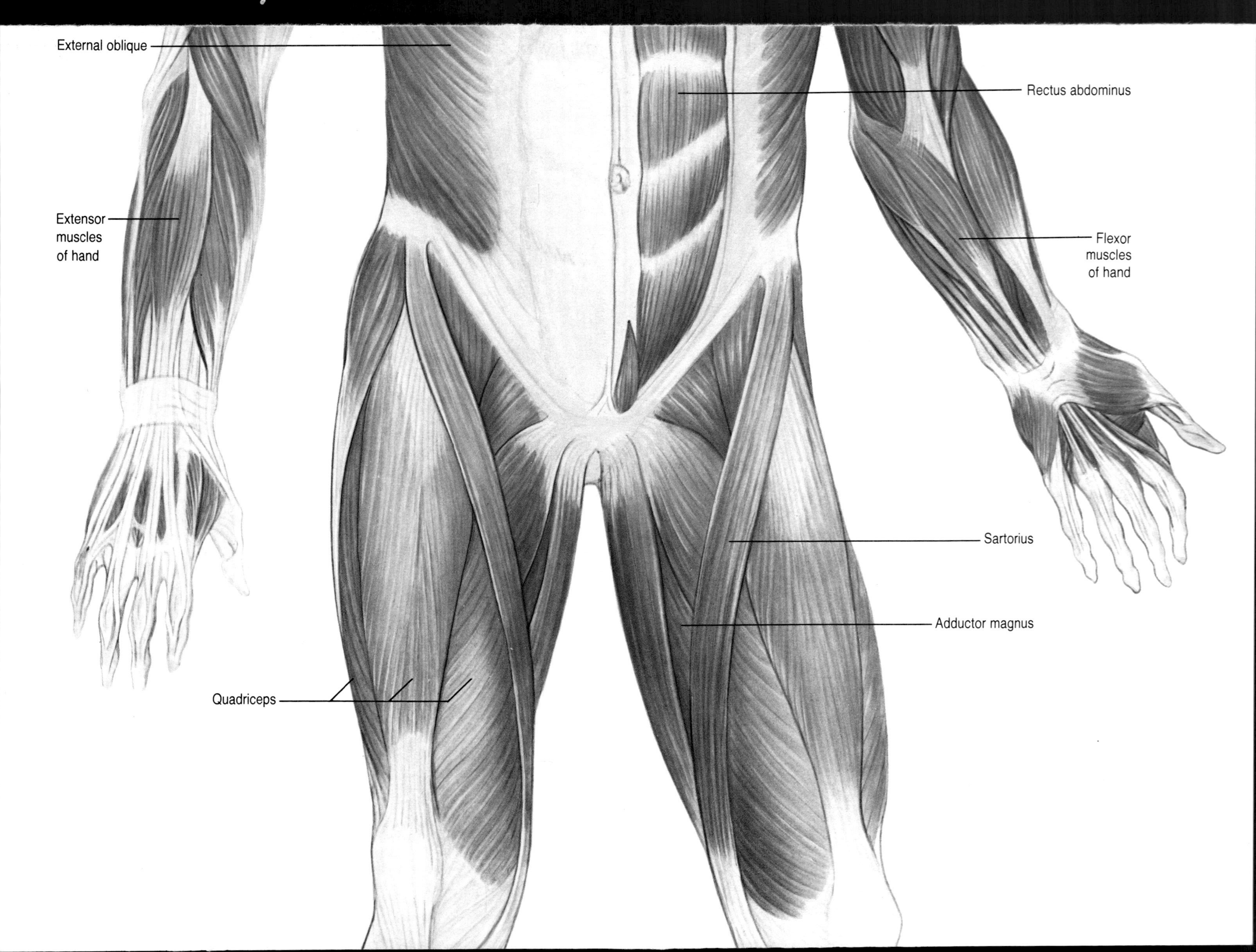
External oblique
Rectus abdominus
Extensor muscles of hand
Flexor muscles of hand
Sartorius
Adductor magnus
Quadriceps

MUSCLES

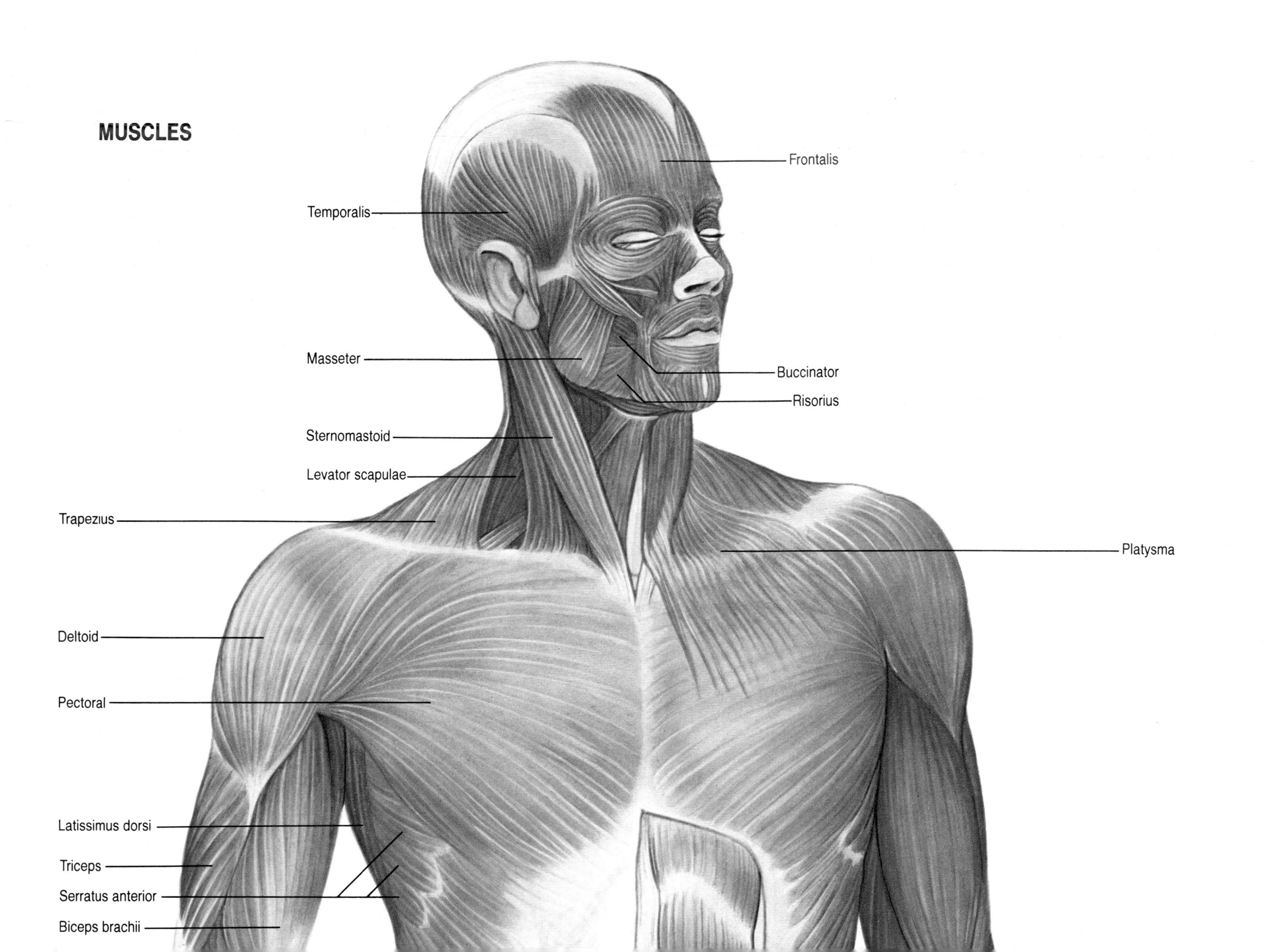
Frontalis
Temporalis
Masseter
Buccinator
Risorius
Sternomastoid
Levator scapulae
Trapezius
Platysma
Deltoid
Pectoral
Latissimus dorsi
Triceps
Serratus anterior
Biceps brachii

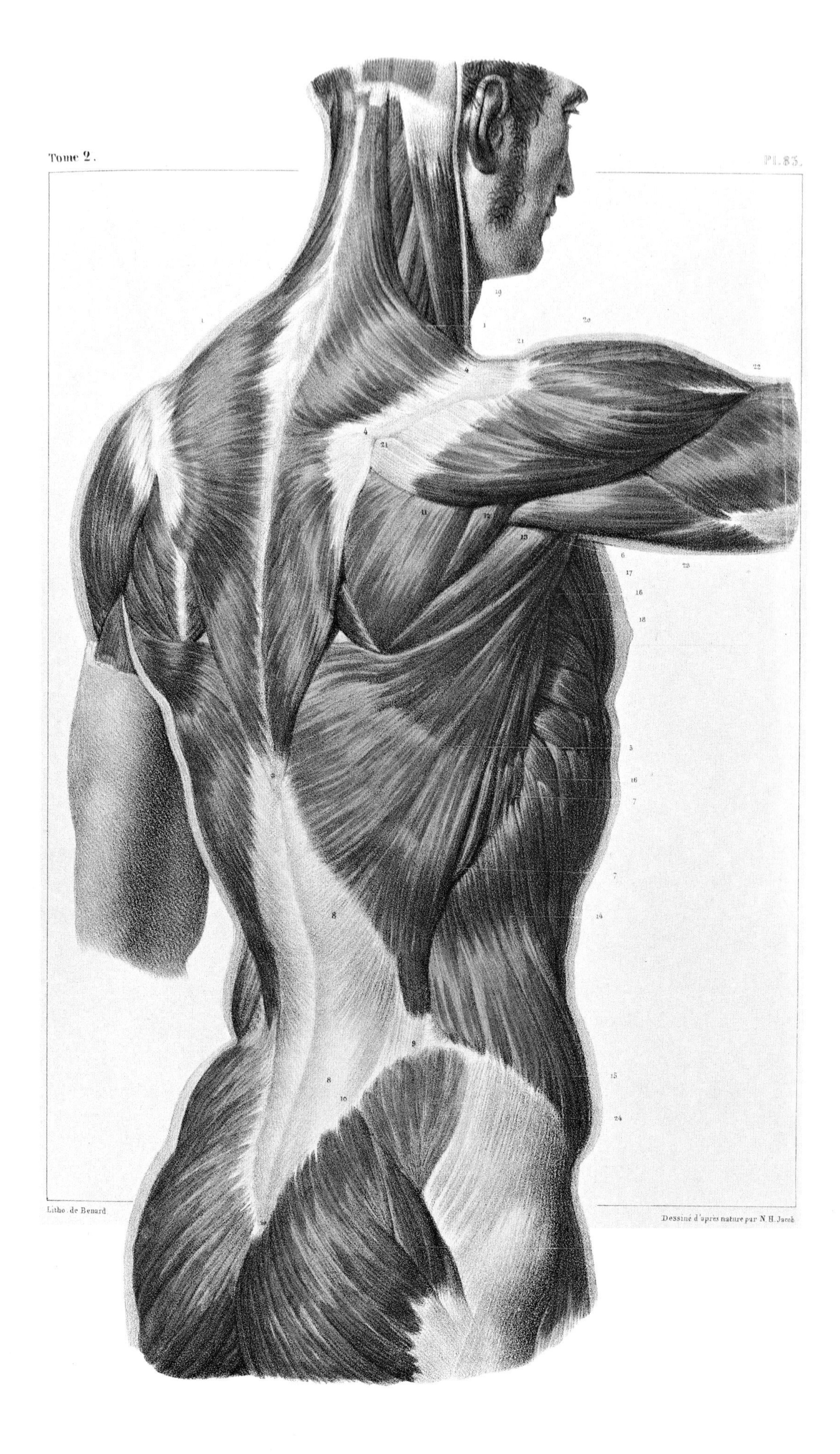

Litho. de Benard

Dessiné d'après nature par N. H. Jacob

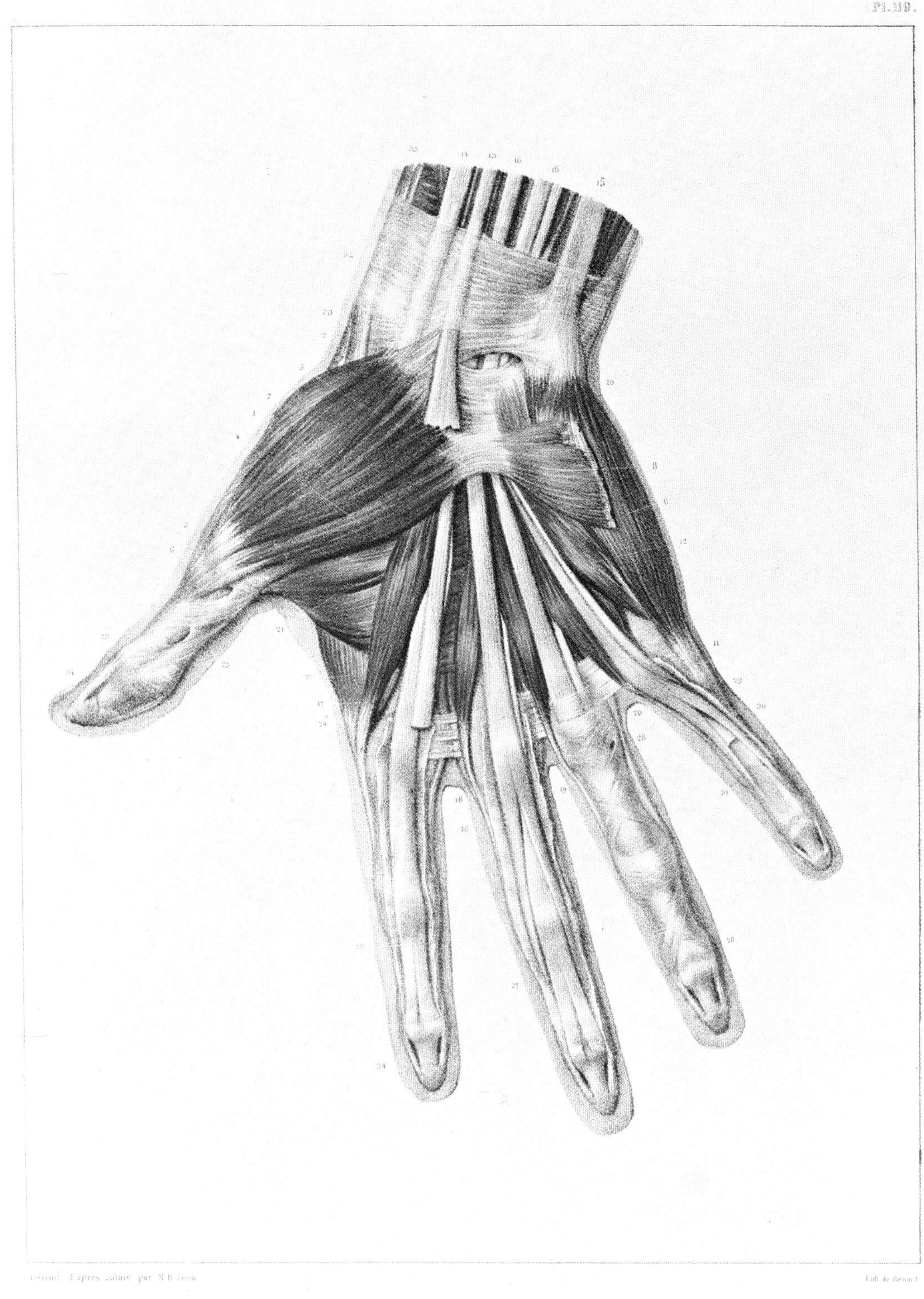

From Traite Complet de L'anatomie de l'homme *by Marc Jean Bourgery (French anatomist) 1831.*

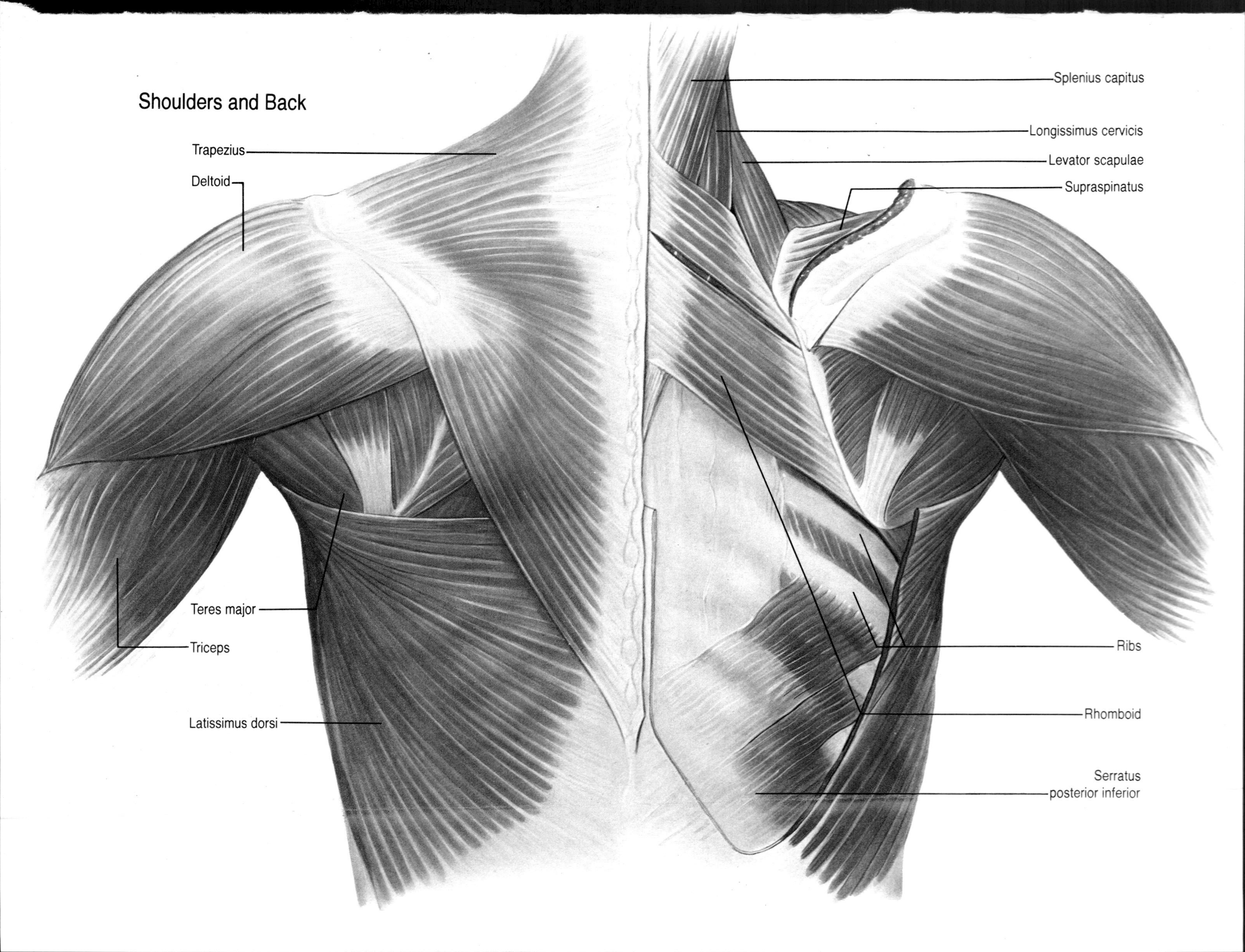
Shoulders and Back
Trapezius
Deltoid
Teres major
Triceps
Latissimus dorsi
Splenius capitus
Longissimus cervicis
Levator scapulae
Supraspinatus
Ribs
Rhomboid
Serratus
posterior inferior

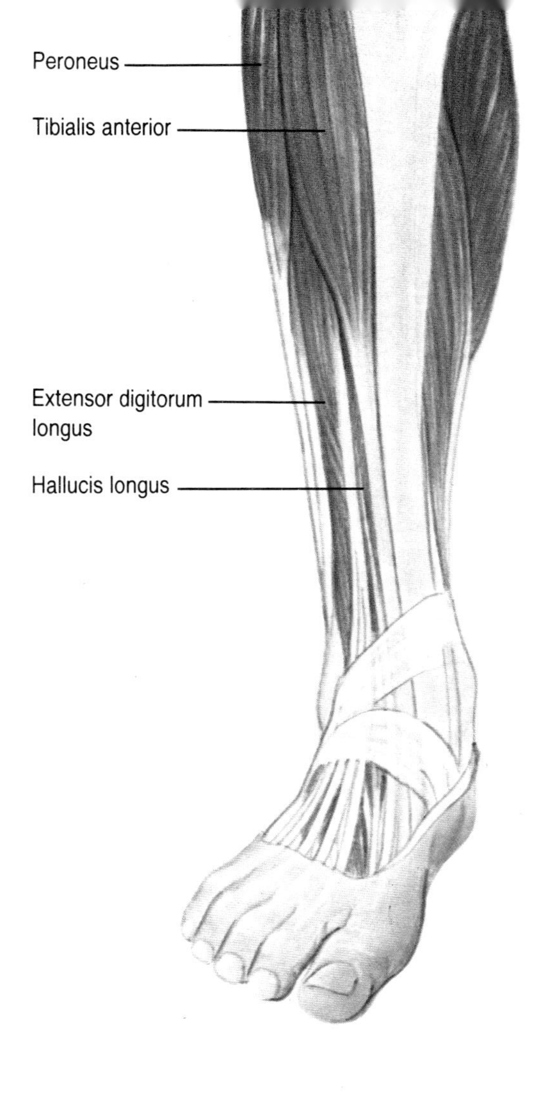
Peroneus
Tibialis anterior
Extensor digitorum longus
Hallucis longus

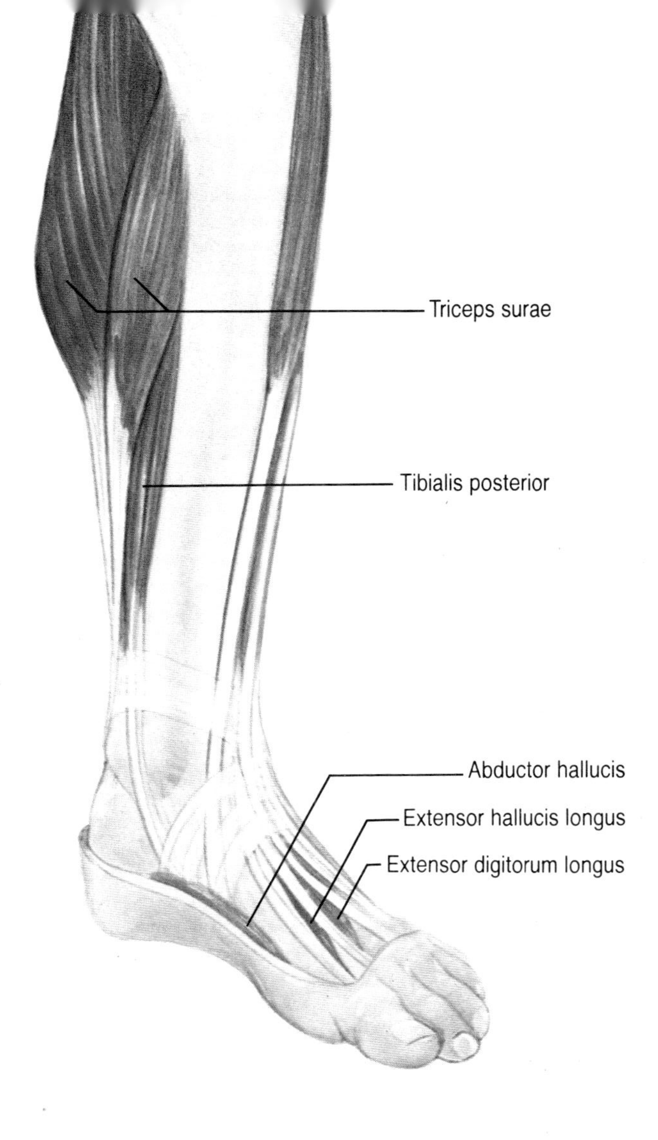
Triceps surae
Tibialis posterior
Abductor hallucis
Extensor hallucis longus
Extensor digitorum longus

The Gait Analysis Laboratory at Boston Children's Hospital Medical Center uses computer technology to study the way people walk. The stick figure below is a computer-generated reading of a normal gait.

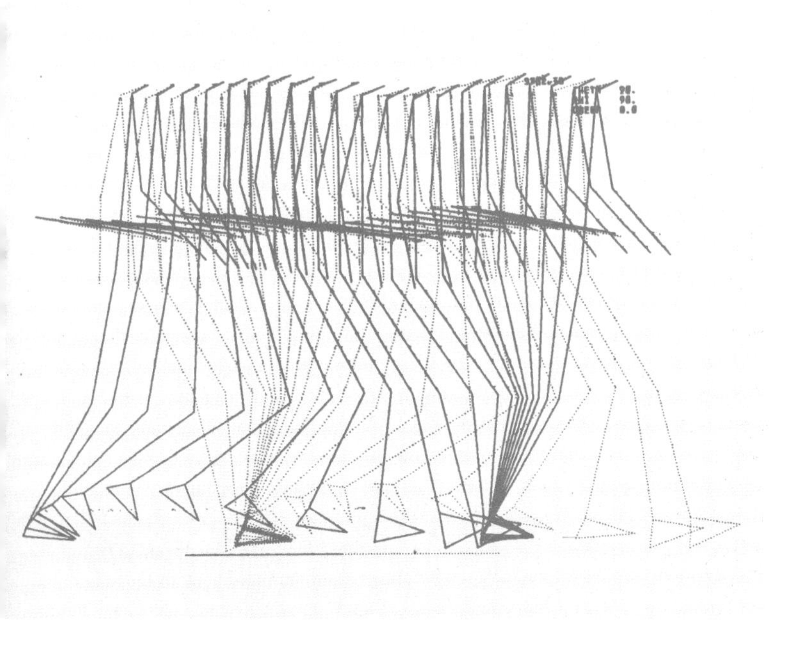

tion of force. The faster the engine contracts, the greater the force that can be produced at the end of the moving lever. The slower the muscular contraction, the less energy required. The basic principle of a lever of the third type is illustrated in the flexion of the arm by the biceps muscle, exerting an upward pulling force on the forearm.

In the human body, speed and range of motion prevail over force. The body best accomplishes tasks involving fast movement with light objects. When great force is demanded during heavy work, the body typically proves inadequate. Man has therefore invented machines and tools, such as the crowbar, that extend his muscular leverage to gain a force advantage.

Work most often involves the use of more than one of the body's lever systems. Even when movement of a single lever does take place, many other parts of the body must be held fast. When force is contingent on speed at the lever's end, other levers league together. Working in sequence, each exerts force as the previous one reaches its peak speed, much like the wind-up before a pitcher releases a baseball. But when many levers engage in a heavy task, like pushing with all one's weight against a door, they function in unison.

Kinesiologists, those who analyze movement, study anatomy and physiology, the science of the function of the body. The combination has long intrigued curious minds. Applying geometry to the task, Aristotle was the first to describe the complex process of walking, a rotary motion. Its intricacies are still not fully understood. From Aristotle's analysis to modern scientific attempts to develop cybernetic prostheses for the crippled, kinesiology encompasses the sum of what is known about human motion.

Essentially, walking is a process that upsets the mechanical stability of the body. Balance is continually lost and regained as each step establishes a new base by alternately moving arm and leg levers forward. The legs generate a rotary motion that propels the body in linear progression. As body weight rolls forth, a leg lever advances, employing the ankle as its pivotal fulcrum. Next, the foot swings forward establishing a new base of support. In this motion, the hip becomes the axis as the other leg lever advances.

Posture is the same in walking as in standing, but the body's center of gravity continually advances, which helps to create a progressive motion that is forcibly channeled in the direction of desired movement. Each leg alternates between a supporting and a swinging role. During the initial supportive phase, the dorsiflexors (the extensor digitorum longus that extends the toes and flexes the foot, the tibialis anterior elevating the foot and the hallucis longus that extends the foot) contract, and the right foot strikes the ground. The gluteus maximus begins to contract, preventing the pelvis from dropping forward as the body's center of gravity advances. The extension of the hip, knee and ankle generates a counter-pressure between the ground and the left foot, and the body begins to move forward.

As the left leg completes its backward push against the ground, it begins to swing forward. This motion would be impossible without the dorsiflexors and the tibialis posterior, responsible for the wear of a shoe heel. They contract, the leg lever shortens and the foot clears the ground. The hip flexor swings the leg forward, with the hamstring checking the advancing momentum. Abductors and adductors, muscles that respectively draw limbs away from and toward the trunk, rotate the pelvis as weight supported by the right foot is thrust forward. The tibialis anterior and peroneus muscles extend the foot, and the tibialis posterior relaxes as knee and ankle flex. As body weight shifts to the left foot, the triceps surae, which stabilizes the leg, the extensor digitorum and the peroneus muscles contract — and the right heel elevates.

Body weight is conveyed from the great toe of the right foot to the heel of the left foot. As its toes strike the ground, two other muscles are summoned into motion. The abductor hallucis and the flexor digitorum brevis, which extend the toes and flex the foot, contract. Kinesiologists are still uncertain of what benefit, if any, these muscles provide. This smooth transference of weight across the foot and toes softens the shock of impact. By the time the great toe of the left foot accepts the weight, the heel of the right foot is already bearing the brunt in a double support system that provides a further safeguard against injury. Step after step, the cycle of muscular contraction and relaxation continues, aided by arms swinging in coordinated movement with legs.

The Laws of Motion

Babies often run before they walk, for they are not yet practiced in negotiating body weight over small and unstable feet. Typically, a baby stands with feet wide apart, ensuring a wide base of support. Leaning forward, his center of gravity advances and, in a natural though erratic sequence of rapid steps, he is able to maintain upright posture while he scampers to a person or piece of furniture for support. When running, between the time the back foot lifts off the ground and the forward foot strikes it, there is momentarily no support. The action is almost entirely propulsive. It takes less energy to maintain a given speed than it does to reach it. In physics, this concept is the first law of motion, developed by seventeenth-century English mathematician, scientist and philosopher Sir Isaac Newton. He maintained that an object, whether at rest or in linear motion, remained in that state unless additional force was applied. Modern physicists call this tendency inertia.

The heavier the object and the faster it moves the more force it takes to overcome inertia and change velocity of movement. Once movement begins, it takes less force to maintain speed than to change it. In the Australian crawl and other arm-over-arm swimming strokes, the successful speed swimmer must execute a second stroke before the forward momentum from the first is lost, due to water resistance. If the glide is maintained until momentum is lost, more energy will have to be expended to overcome inertia on each stroke.

In football, if a light halfback must block a larger lineman, his main ally is Newton's second law of motion. To stop his opponent at the line of scrimmage, the halfback must develop sufficient speed and muster enough momentum to overcome his adversary's greater weight advantage. Although Newton did not have the halfback's plight in mind, he sounds remarkably like a football coach warning his players about the risk of injury in blocking a lineman. "If a body impinge upon another, and by its force change the motion of the other," Newton wrote, "that body also (because of the equality of the mutual pressure) will undergo an equal change, in its own motion, towards the contrary part." In other words, a hard-hitting block can exert as much force upon the halfback as on his opponent.

When a runner sprints forward, he pushes back and down on the ground with his back leg. In so doing, he moves forward and upward with the same rate of force. Every action has an equal and opposite reaction. This action-reaction response composes Newton's third law of motion. The principle applies to swimming, too. If a swimmer pushes off with a force of twenty-five pounds, the resulting takeoff propels him forward with equal force. A diver who pushes down against the surface of a springboard is pushed up by the board; if he pushes backward, the board pushes him forward. This law enables a diver to initiate both twist and spin dives before ever leaving the board. Once a diver is in the air, if he moves one part of his body in one direction, the rest of his body reacts in a motion complementary to his momentum. Arms swung to the right and across his chest will twist the body with equal force to the left.

Time exposure photography, capable of stopping a swimmer's stroke to each fiftieth of a second, captures the strenuous effort inherent even in graceful glides. Analyzing the movement of light attached to a swimmer's wrist can help an athlete improve speed and precision. The same technique freezes a diver's faultless tuck and illustrates the action-reaction principle that makes springboard diving possible.

The center of gravity, the imaginary point where body weight is equally distributed, is the key point around which a diver spins. When bent over in the jackknife or pike position, the center of gravity can lie outside the body, in the space between trunk and legs. The tightrope walker whose fame rests on the refusal of a safety net must always keep his constantly shifting center of gravity above the tightrope. The circus performer's body is like a figure constructed of building blocks. Only blocks perched precisely above one another will stand. But if even one block is moved, drastically shifting the center of gravity, the blocks collapse. When his arms fall out of line, gravitational force pulls him downward. The center of gravity shifts and the body falls out of balance. Pressure is exerted unevenly and the tension of opposing ligaments becomes unequal. Balance can be regained only by shifting body weight, thus restoring the center of gravity above the base. Once achieved, total body balance can be maintained as each part is centered over the part directly below it. Good posture is the position in which the center of gravity of each body segment is centered over its supporting base. The body is then in equilibrium and muscular effort on one side is equal to that

Guillaume Duchenne

Electricity's Healing Touch

Guillaume Benjamin Amand Duchenne secured electrodes to nerves in the base of the cadaver's neck. Watching intently, he switched on the current, which jolted the lifeless body with electricity. Duchenne's eyes grew wide when the dead man drew a breath of air.

Duchenne's research on electrical stimulation of human muscles, living and dead, culminated in 1867 with the publication of *Physiology of Motion.* The 832-page book would revise classical observations of anatomy and physiology of muscles set down by the great scientific minds of the ages.

The experiments he performed established Duchenne as a pioneer in neurology and what is now broadly known as kinesiology, the study of motion of the human body.

Through his perseverance and ingenuity, he catalogued the action of muscles more completely than anyone had done before.

As a student, however, he apparently showed no particular interests or abilities during his years of medical school in Paris. In 1831, at the age of twenty-five, he started a private practice in general medicine in Boulogne, the seaport city of his birth. He soon became fascinated by the idea of

using electrical current to treat rheumatism, paralysis and other afflictions. Duchenne discovered a way to introduce current painlessly into the body by applying electrodes against the skin. Others in this bold era of experimentation were inserting platinum needles through the skin, which not only caused great pain but killed surrounding tissue.

He found that stimulation of certain nerves prompted muscular response. Lured back to Paris, the center of French medical thought, Duchenne continued experiments that soon won him the attention of colleagues in the large hospitals. He constructed his own apparatus which permitted unprecedented accuracy in isolating and stimulating individual muscles. Working with the body of a patient who had died only minutes before, he "applied a very intense high-frequency induction current to the phrenic nerve [the nerve controlling the diaphragm]" and "observed . . . fairly distinct inspirations."

Duchenne's experiments, which he also conducted on living subjects, demonstrated that isolated contraction of the diaphragm lifted and expanded the ribs. His innovative research made clear for the first time the diaphragm's specific role in breathing. With further study, Duchenne was able to catalogue the action and use of other muscles involved in breathing and explain diaphragmatic paralysis.

In *Physiology of Motion,* Duchenne described movements of the entire muscular system. He thought his book "essentially practical" for identifying and treating diseases of the muscles. He wrote, "Without the physiologic principles it teaches, it would be impossible to make a precise diagnosis of paralysis or partial atrophies." The principles he propounded did lead to better treatment.

In this detail from Hammerstein's Garden, *painted in 1901, William Glackens captured the thrill of human motion perfected. The tightrope walker's skill rests on her ability to achieve precise body balance.*

on the other. The tightrope walker uses gravity to advantage in keeping the weight-bearing joints of the body in strict alignment.

Erect posture, so important to the high-wire performer, is, nevertheless, a subjective position. Aesthetic and cultural standards establish some criteria for what may be considered good posture, but there is no objective, practical method of judging posture. The rigid military posture is not necessarily the best stance because it requires more energy to maintain than relaxed posture. Kinesiologist Eleanor Metheny wrote, "For each person, the best posture is that in which the body segments are balanced in the position of least strain and maximum support."

The maintenance of posture seems to be coordinated by five separate reflexes that lie in the midbrain. Stimuli received from these sources appear to bring appropriate muscles into action to correct displacements. All muscles develop a certain degree of vigor and tension called muscle tone. When the tone of two antagonistic muscles becomes unbalanced, affected body parts may deviate from normal position. A waiter who daily carries heavy trays of dishes in front of him might develop lordosis, a postural defect caused by habitually assuming a swayback stance. If such posture becomes second nature, the muscles of the lower back will gradually shorten and the abdominal muscles will elongate. Often, the sufferer has only to practice a more erect posture until it becomes habit. Proper stretching exercises can also help correct posture by elongating back muscles and shortening abdominal muscles.

Although science has made great strides in analyzing the body's 600 engines, even simple voluntary movements — picking up a book or turning a page — continue to puzzle researchers. That progressive movements constituting behavior spring from an incessant interaction between sensory feedback and muscle actions is known. This interplay, sensory-motor integration, occurs between the spinal cord and the part of the brain that specializes in coordinating sensation and response. Man is deaf to this chatter that engenders movement. Still, conscious and unconscious impulses wed in an often indistinguishable bond that consummates human movement.

Chapter 4

Circuits for Movement

Like a diaphanous seedling reaching toward a claret sky, a motor nerve fiber permeates ribbons of skeletal muscle. Currents of movement sweep through the circuitry of the nervous system, culminating at clusters of buds called motor end plates. A chemical surge spills from the buds, galvanizing muscles into motion.

Only through movement does man declare his being. Unable to move, he would be but an effigy of himself. Thought and feeling would remain forever locked inside the mind were it not for the movements of the lips and tongue in speech, or of the hand inscribing words on paper. Every movement, simple or elaborate, is a sequence of skeletal muscle contractions and relaxations woven seamlessly by means of a complex interplay between the body's muscles and the nervous system. An elementary motion like picking up a book from a table demands the perfectly synchronized interaction of many muscles. Muscles that flex the arm at the elbow must relax to allow the forearm to extend. Shoulder muscles must stabilize to support first the weight of the extended arm, then the book. As the body pitches forward to reach, the muscles of the trunk and legs must compensate for the shifting center of gravity.

A complete understanding of how the body produces such voluntary movements has eluded centuries of scientific investigation. The mechanisms are labyrinthian. Every level of the nervous system contributes uniquely to movement, yet, through multitudinous feedback loops, also has the capacity to influence the nature of messages arising at every other level.

Nearly all movement involves both conscious and unconscious elements. The decision to pick up the book is willed, but most of the muscle activity involved in the action takes place unconsciously, automatically. "The mind orders a particular movement," said the distinguished British physiologist Edgar Adrian, "but leaves its execution to the lower levels of the nervous system."

The command to move a skeletal muscle is initiated in the brain and conveyed over millions of neurons, the special cells of the nervous system, to the muscle. Projecting from the cell body is a long fiber, the axon, and one or more shorter ex-

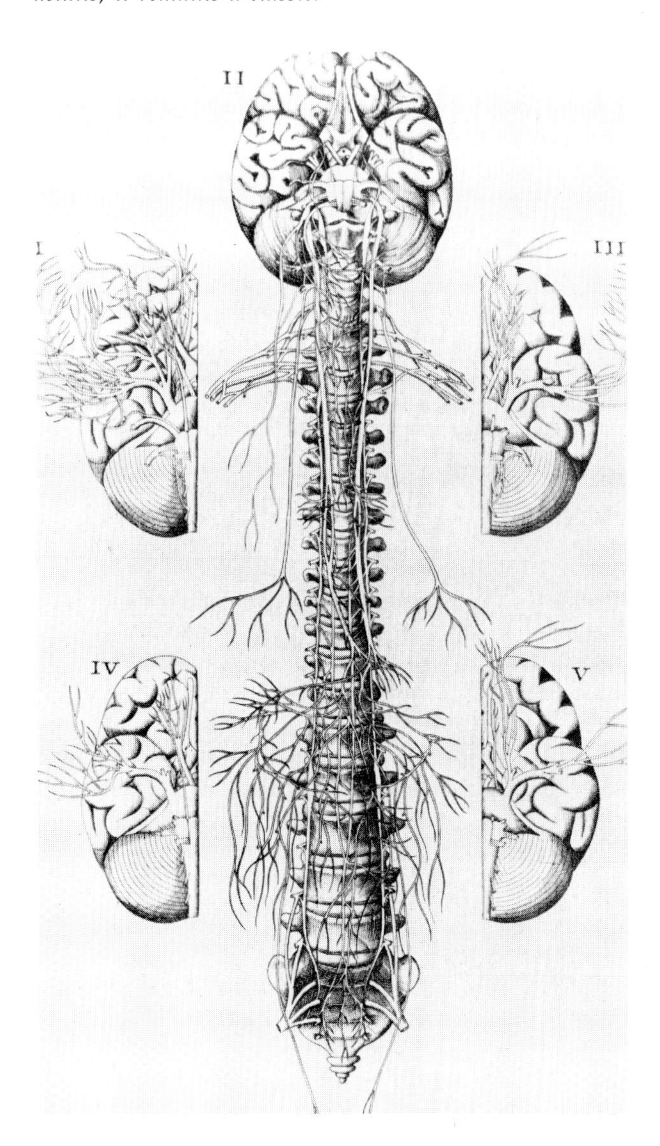

Renaissance anatomist Eustachius depicted the base of the brain and the sympathetic nervous system in this copper plate, engraved in 1552. Although incorrect in certain details, it remains a classic.

tentions called dendrites. Neurons vary widely in size and in the arrangement of their axons and dendrites. Those situated in the same region or performing similar functions often resemble each other structurally, which permits them to respond to nerve impulses in a similar fashion. The majority of neurons are said to be multipolar, meaning that they have one axon and several dendrites. A multipolar neuron transmits either motor or sensory nerve impulses. Rarest is the bipolar neuron, with one axon and one dendrite. Found in the eye, nose and ear, they perform sensory functions. Unipolar neurons in the spinal cord and its nerves have two projections that fuse together just beyond the cell body, and then split apart to form a distinct axon and dendrite.

Neurons communicate by means of neurotransmitters, chemical substances that produce electrical impulses. At the junctions between adjacent neurons lie synapses, incredibly small gaps, about one-millionth of an inch wide. In most neurons, sacs in the end of the axon burst open, spilling neurotransmitters across the synapse. The neurotransmitters lock onto special sites on the cell body, dendrites or axon of the receiving cell. In rare instances, nerve impulses may pass from dendrite to dendrite. Some neurotransmitters are excitatory, tending to pass on an electrical impulse by increasing the positive electrical charge within the neuron. Others are inhibitory, tending to impede the impulse by increasing the negative electrical charge within the neuron. A single neuron may have hundreds or even thousands of synaptic contacts on its surface. The net effect of this chemical antagonism determines how a neuron will respond.

Sensory neurons carry nerve impulses from all parts of the body to the spinal cord or the brain. Motor neurons transmit nerve impulses away from the spinal cord or brain to a muscle or gland. Interneurons, in the brain and spinal cord, are liaisons between sensory and motor neurons, conveying signals between the two. Impulses from motor neurons travel down the neuronal axon to the muscle fibers, where they trigger the release of the chemical compound acetylcholine, setting off a chain of events that ends in a skeletal muscle contraction. Within a second, millions

Firing the signals that kindle movement, neurons are the cells of the nervous system. All are united by an electrochemical language that travels various pathways between brain and muscle.

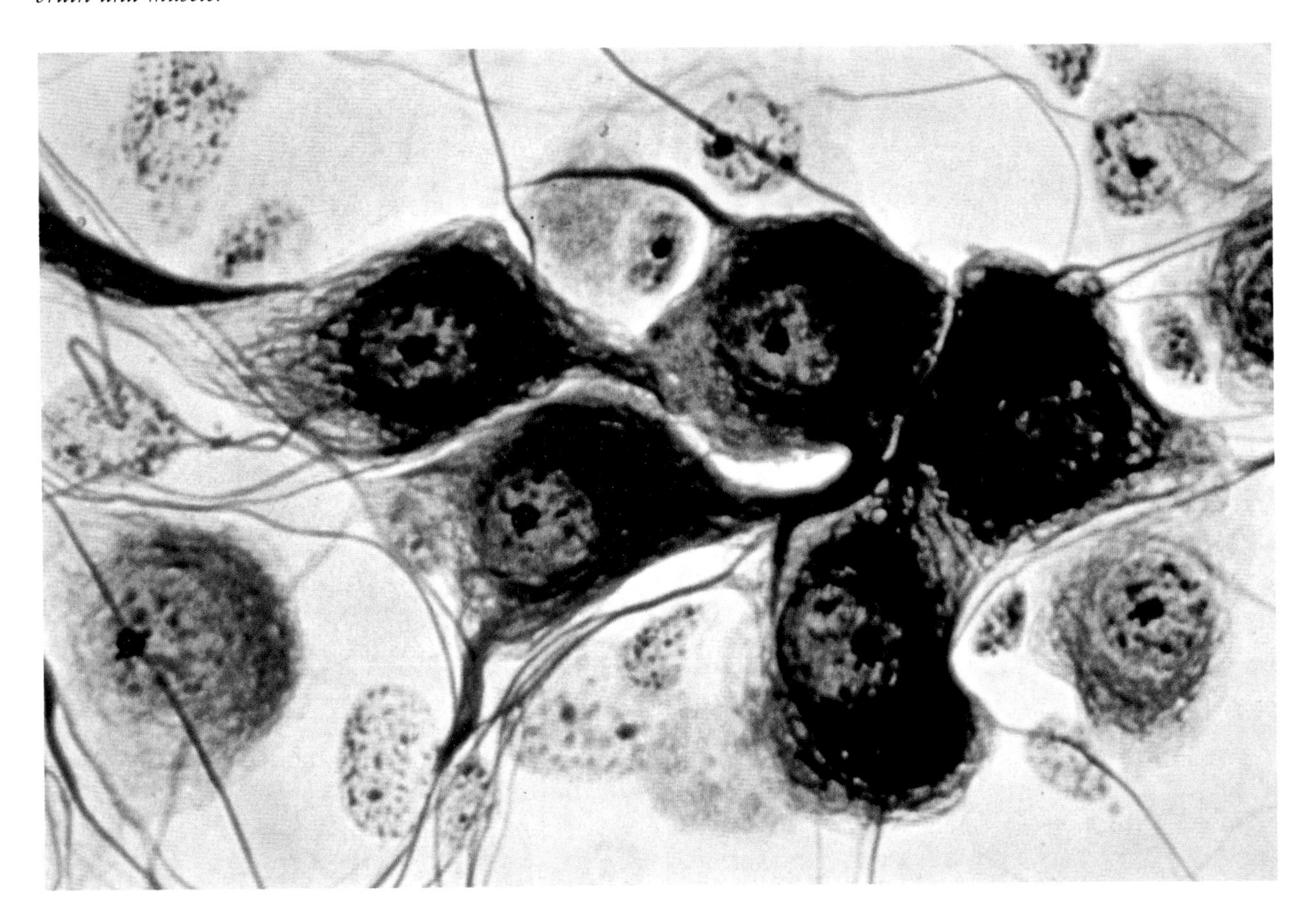

of impulses reach the motor neurons, some sent from various parts of the brain and spinal cord, others from special sense organs located in joints, ligaments, tendons and the muscles themselves.

"An Insuperable Objection"

The seeds of movement, however, are sown by the brain. What is now known as the primary motor cortex, a region of the brain's wrinkled surface spanning both cerebral hemispheres, was discovered little more than a century ago. By experimenting on living dogs, German physicians Eduard Hitzig and Gustav Fritsch found that electrical stimulation in this particular region caused the animals' bodies to move. It was a momentous but upsetting discovery. Traditionally, only the higher mental functions were thought to be the province of the cerebral cortex. "There seems to be an insuperable objection to the notion that the cerebral hemispheres are for movements," remarked renowned British neurologist John Hughlings Jackson in 1870. He had already proposed the idea from observing victims of stroke and epilepsy. "The reason, I suppose, is that the convolutions are considered to be not for movements but for ideas."

In experiments that would revolutionize concepts of the brain, researchers in Europe and America began to plot the cortex of higher primates. In the first two decades of the twentieth century, Hughlings Jackson's fellow countryman, Sir Charles Sherrington electrically stimulated the primary motor areas of chimpanzees, gorillas and orangutans. He found that stimulation of the same point often elicited different responses in different animals. The higher the animal on the evolutionary ladder, he discovered, the more sophisticated and "fractionalized" the primary motor cortex. Stimulation of a certain cortical point that might cause both sides of a dog's jaw to

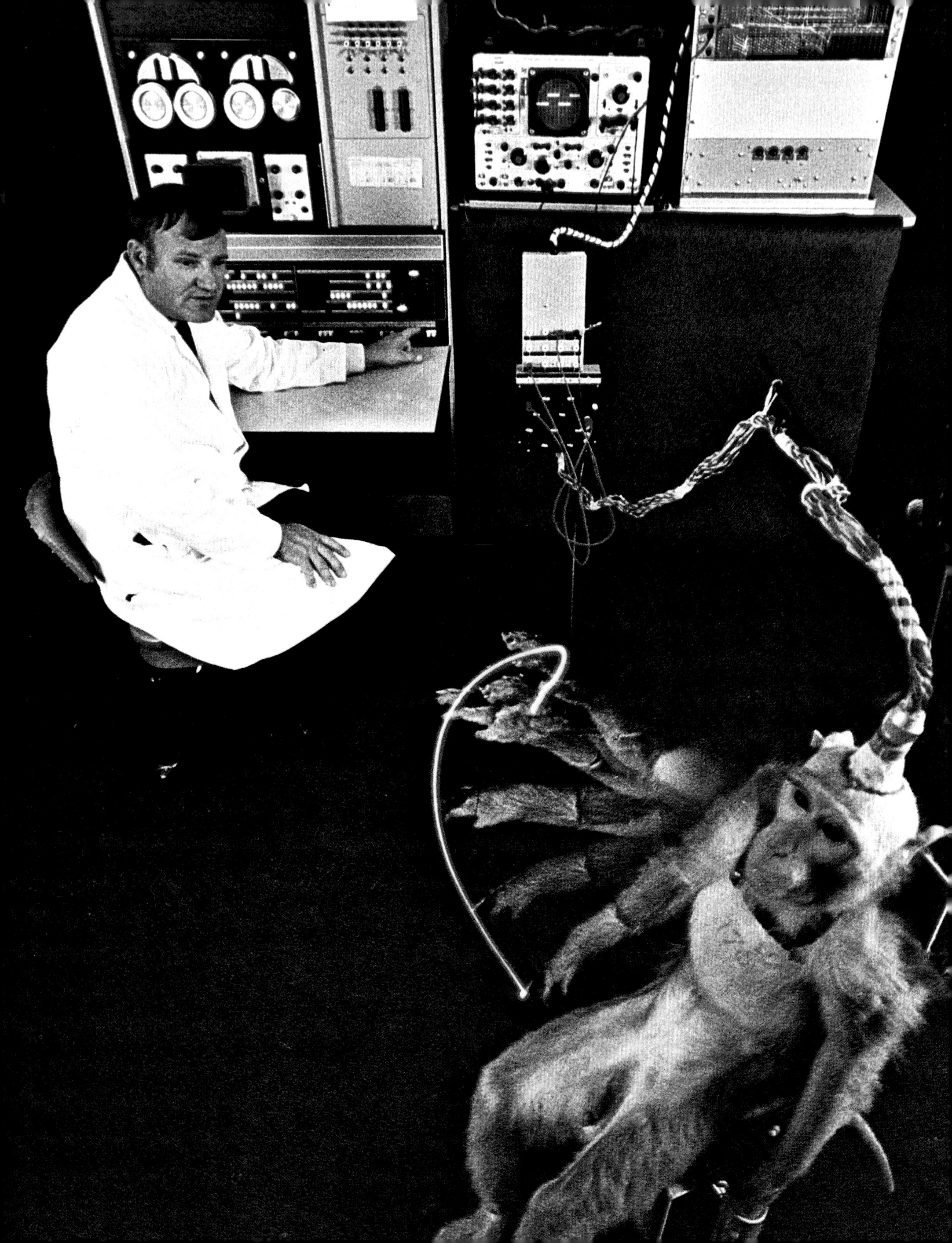

Electrodes planted in a monkey's brain pass minute currents into the motor cortex, a major center of movement, opposite. The scientist employs a computer to analyze muscular responses to cortical stimulation.

The cerebral cortex also shelters the secondary motor region, thought to be crucial to intricate movements. Sensations registered in the somatic sensory cortex supply feedback vital to any willed action.

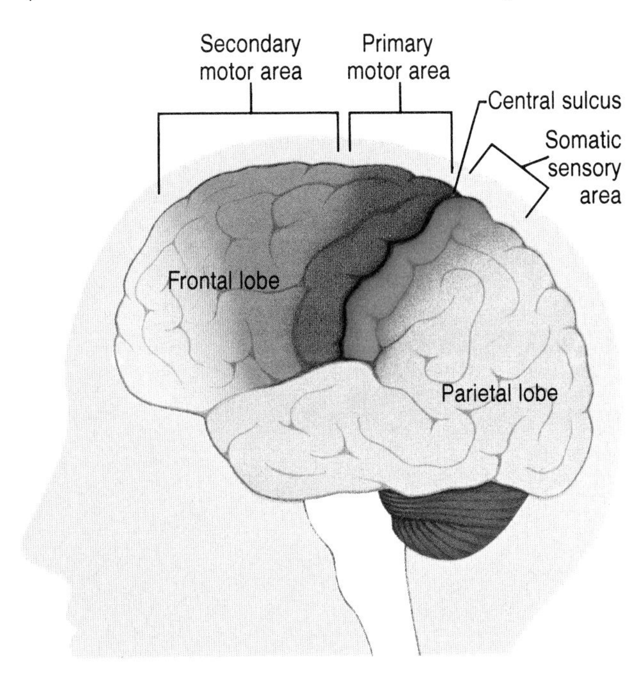

open might provoke movement in only one side of an ape's jaw. Even in animals belonging to the same species no two cortical surfaces were exactly alike. Nor would the same movement occur in an individual animal every time a particular point on the primary motor cortex was stimulated. Greater fractionalization, it seemed, provided finer control of movement in higher animals.

The primary motor cortex lies in a ridge just in front of the central sulcus, a deep furrow that runs vertically along each of the cerebral hemispheres, separating the frontal and parietal lobes. A student of Sherrington's, Canadian neurologist and surgeon Wilder Penfield produced a comprehensive map of the human motor cortex in 1947. By electrically stimulating cortical points on epileptic patients under local anesthesia, he noted which muscles moved. Known as the motor homunculus, the map revealed that all parts of the body are not equally represented in the primary motor cortex. Where body movements are more precise, the parts are allotted a larger representation in Penfield's schematic drawing. Muscles of the hand and mouth together occupy almost two-thirds of the primary motor cortex.

Here, the body's reflection is topsy-turvy. Neurons are organized in an inverse relationship to the parts of the body they control. Those that produce movement of the feet are at the top of the primary motor cortex, followed by those controlling the leg, torso, arm, hand, fingers and, finally, at the very bottom, the neck and head. Modern-day scientists have found that the neurons are also arranged in perpendicular columns. Each column contains tens of thousands of neurons and seems to perform a specific motor function. One column may stimulate a particular muscle, while another may activate groups of muscles that work together. Since each column acts as a unit, the time spent in relaying electrical impulses is minimized. A patch of cortex directly in front of the primary motor area also houses neurons involved in movement. Researchers think this region, the secondary motor area, is crucial to speech and intricately coordinated movements such as those the hands perform.

Electrical impulses from many regions of the brain feed into the motor areas. Before the nervous system can orchestrate a coordinated movement, it must collect and integrate all sensory messages. One of the most important sources of information for movement is the somatic sensory cortex, lying on the other side of the central sulcus directly behind the primary motor cortex in the parietal lobe. Here, the brain registers sensations. Interplay between the senses and movement is continuous and elaborate. Sight, sound, smell, pressure and pain are naturally important but equally so are messages from the muscles and related tissues, which relay information about the angles and positions of the joints, the length and tension of muscles, even the speed of movements. The motions that form a coordinated movement arise from this continual feedback between sensations and muscle actions.

For decades, the motor cortex has been viewed as the highest command center for movement. Newer evidence challenges this concept, ascribing the initiation and control of movements to structures in the lower brain. During the early 1970s, scientists discovered that the cerebellum and a group of nerve clusters known as the basal ganglia, two of the brain's oldest evolutionary components, may act at a more abstract level than the motor cortex in the chain of motor com-

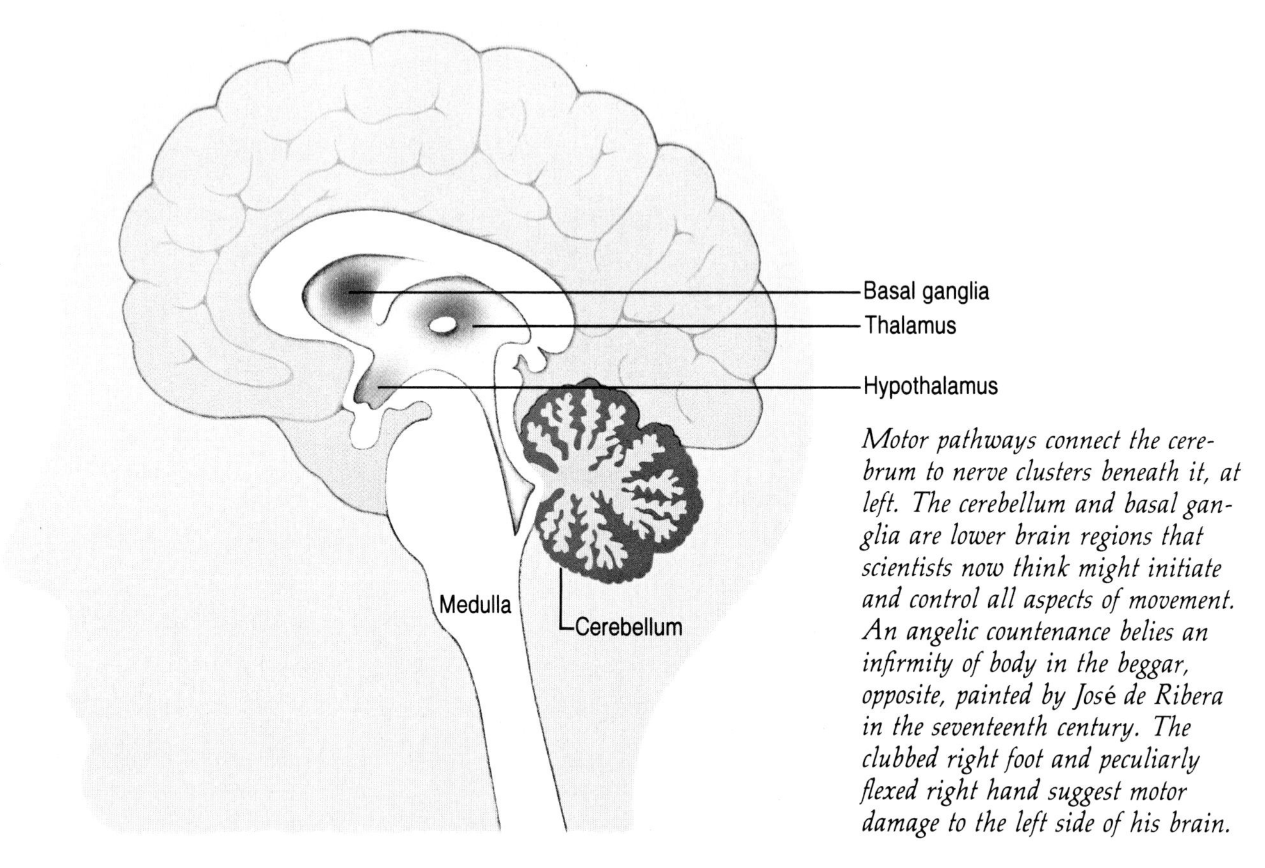

Motor pathways connect the cerebrum to nerve clusters beneath it, at left. The cerebellum and basal ganglia are lower brain regions that scientists now think might initiate and control all aspects of movement. An angelic countenance belies an infirmity of body in the beggar, opposite, painted by José de Ribera in the seventeenth century. The clubbed right foot and peculiarly flexed right hand suggest motor damage to the left side of his brain.

mand. These lower brain structures may initiate messages for movement and send them to the motor cortex, which relays them through nerve pathways to the spinal cord and, finally, to the skeletal muscles. Thus, the motor cortex may refine muscle actions rather than will them.

Behind the cerebral hemispheres and attached to the brainstem, the brain's primitive central core, lies the cerebellum. As in the cerebrum, a deep groove divides the cerebellum into two lobes. Its wrinkled surface encloses a multitude of folds. The cerebellum receives a rich variety of sensory information through both its neuronal connections with the somatic sensory cortex and directly from peripheral parts of the body. Most important, it coordinates muscle actions, controls reflexes crucial to posture and helps maintain the body's equilibrium. The cerebellum is crucial in the control of skilled and rapid muscular patterns, the kind required for talking, typing or playing a musical instrument. It instantaneously compares information from various areas of the brain about what the muscles should be doing to reports from sensory receptors in the muscles, tendons and joints that track what the muscles are actually doing, and then corrects any discrepancy. The cerebellum adjusts muscle movements to a given task so they are not too abrupt, too weak, too far or too near. When performing, the ballerina exemplifies perfect cerebellar functioning. Injury or disease to her cerebellum would mean the end of a career. The act of putting on her ballet slippers would alone become difficult. Reaching for them, she might overshoot the mark. Tying them, her hands might shake violently. Overshooting, known as ataxia, is often seen in the elderly and sometimes accompanies cerebral palsy, a disorder caused by brain damage during fetal development or birth.

Learning any activity, like swimming or tennis, first requires concentration on every movement. Once the activity has been learned, such deliberate attention becomes unnecessary. Scientists speculate that different motor systems might be involved in the initial learning and in the performance of a skill that has been mastered, with the cerebellum playing a key role in the latter.

Because the cerebellum plays a special part in the automatic control of movements, some scientists see it as a counterpart to the basal ganglia, a group of nerve clusters in the cerebrum's interior. The cerebellum is especially important in the execution of movements once they have begun, while the main function of the basal ganglia ap-

DAMIHI ELIMO
PROPTER
REM DEI

pears to be the initiation of voluntary movements. In birds and lower animals with a relatively small cerebral cortex, they are the highest centers for movement. In human beings, the basal ganglia, like the cerebellum, are part of a feedback loop to the motor cortex. Cerebellar and basal ganglia links to the motor cortex perform different functions. The cerebellum is connected with the body's sensory receptors to detect and correct discrepancies between an intended action and actual muscle movements. The basal ganglia, however, receive most of their information from the cerebral cortex. The role differences result in different symptoms in patients with diseases affecting the two structures.

Victims of Parkinson's disease, a basal ganglia disorder, suffer from difficulty in initiating movements, tremors, muscle rigidity and loss of facial expression. Described clinically in 1817 by English physician James Parkinson, the disease is also called paralysis agitans, or "shaking palsy." Surgeons can sometimes relieve symptoms by destroying parts of the basal ganglia or portions of the thalamus, an oval mass in the center of the brain that feeds messages from the basal ganglia to the motor cortex. About 75 percent of patients improve when given the drug L-dopa.

A Dancing Delirium

Chorea, a potentially more serious disease of the basal ganglia, induces jerky, spasmodic, involuntary movements resembling a dance. One form strikes children suffering rheumatic fever that has spread to the brain. It is sometimes called St. Vitus's dance, after the widespread fad of dancing mania that swept through Europe during the Middle Ages. Ranting, writhing hordes congregated in the streets and churchyards of Germany, Italy, France and the Low Countries, creating spectacles like this one described in 1374: "They

Dancing manias swept through Europe during the Dark Ages. The writhings, termed St. Vitus's dance, recall the spasms of Sydenham's chorea, a neuromuscular disease that sometimes goes by the same name.

formed circles hand in hand, and appearing to have lost all control over their senses, continued dancing . . . for hours together, in wild delirium, until at length they fell to the ground in a state of exhaustion." St. Vitus's dance lured both the young and the old, but most who joined in the epidemic frenzy were zealots between the ages of fifteen and twenty-five, driven to hysteria by medieval religious fervor. Many sought relief by making a pilgrimage to one of Europe's numerous shrines to St. Vitus, a child saint who died in the fourth century. Thomas Sydenham, a well-known seventeenth-century physician and friend to the philosopher John Locke, described and rather inexplicably named the neuromuscular disease after these dancing manias in 1686. Among Sydenham's beliefs was the notion that epidemics followed some sort of seasonal or yearly cycle. Today, the disease is more often referred to as Sydenham's chorea.

Huntington's chorea is a rare but lethal progressive hereditary disorder that strikes between the ages of thirty-five and fifty. Seldom do victims live more than fifteen years after the onset of the disease. The mental deterioration that accompanies Huntington's chorea can be severe.

The portion of the brain between the cerebrum and the midbrain, known as the diencephalon, contains several nerve centers that also contribute to movement. In the diencephalon, the thalamus relays motor and sensory messages to the cortex. It passes relatively mild sensations on to the somatic sensory area but deals immediately, through reflex action, with strong, crude messages — those of pain, pressure, rough touch and extreme heat or cold. With the cortex, the thalamus also controls facial muscles. Consequently, people who have suffered injury or disease to the motor cortex are still able to register strong emotions on their faces.

Beneath the thalamus is a diminutive globe of nerve cells called the hypothalamus. Regulator of many of the body's basic automatic functions — hormone secretion, body temperature, thirst and appetite — the hypothalamus is also the birthplace of emotions, containing centers for pain and pleasure. It influences movements associated with feelings, from the paralysis that can accompany intense fear or shock to the seemingly uncontrollable motions generated by rage.

The brain's central stalk cradles the brainstem, a control center for breathing, heartbeat and blood pressure. The medulla, at the brainstem's base, is actually a bulbous extension of the spinal cord, becoming a crossroads for one of the two major transmission networks linking brain and skeletal muscles. Beds of nerve fibers swell into twin pyramids on the surface of the medulla, then descend to the spinal cord in nerve tracts known as the pyramidal system.

The pyramidal system is an important mediator of fine, skilled movements. Sometimes called nerve fibers, axons in the pyramidal tracts have their cell bodies in the cortex. About 60 percent come from the motor areas and 40 percent from the somatic sensory region. Pyramidal axons are the body's longest, some stretching two feet. As they descend from the cortex, before reaching the

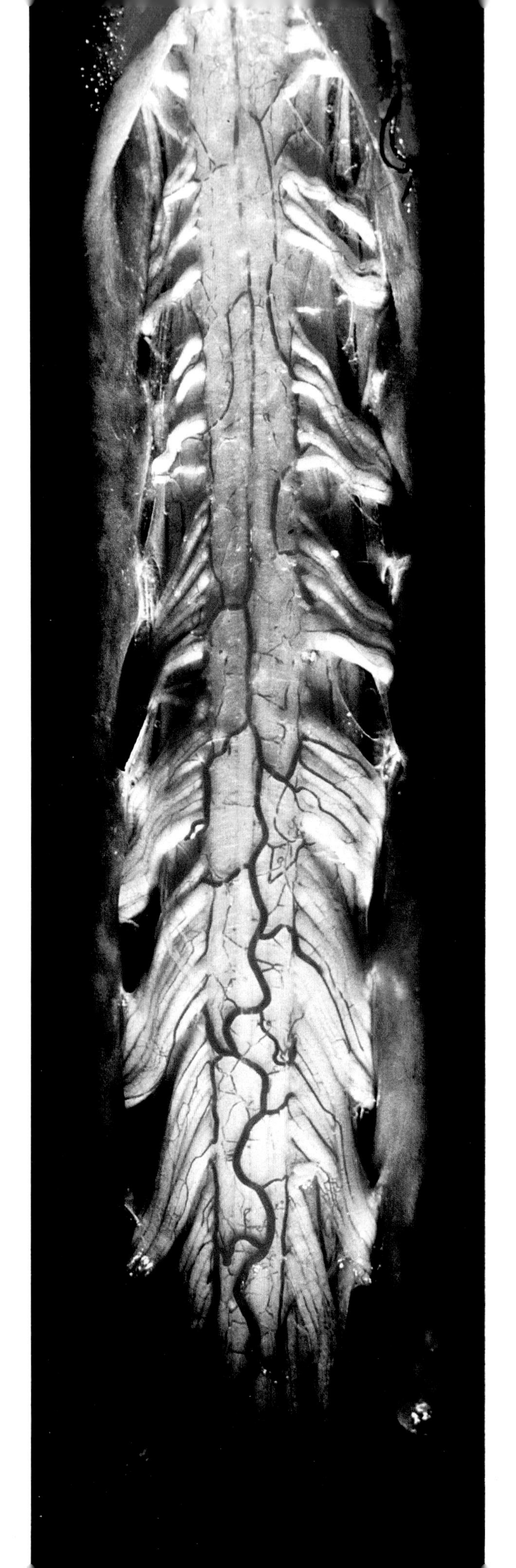

Sensory nerve fibers from spinal nerves feed into the spinal cord, shown here from behind. A mere half-inch wide, the cord houses millions of neurons, many of them from the motor pathways.

medulla, some axons send branches to the cranial nerves supplying muscles that control movements of the face, throat, eyes, glands, upper back and neck. Other pyramidal axons send off collateral branches to various lower brain structures, creating a feedback loop that informs the lower brain about what the cranial and spinal motor nerves are doing. These messages, in turn, are relayed back to the motor areas of the cortex to influence movements.

Inside the medulla, an estimated 80 percent of the pyramidal axons cross over, so that skeletal muscles on the right side of the body are controlled primarily by neurons in the left side of the brain, and vice versa. The axons course down through the spinal cord, most of them terminating at various levels of the cord on interneurons, which in turn fire motor neurons that supply muscle fibers. About 10 percent of all pyramidal axons terminate directly on motor neurons, providing a direct path for quick muscle activation.

At every step along the descent from brain to muscle, nerve impulses from the pyramidal system diverge. A single neuron in the motor cortex can influence numerous spinal cord interneurons, each of which controls many muscle fibers. By regulating the degree of divergence, the nervous system can vary the precision of muscular control. The torso and other parts of the body that have a limited precision of motion are supplied by relatively few pyramidal neurons — perhaps 50,000. Hand muscles, which perform extremely delicate movements, are driven by impulses from roughly 200,000 pyramidal neurons.

The second major transmission network in the control of movement is the extrapyramidal system. Because it produces contractions of groups of muscles either simultaneously or in sequence, the extrapyramidal system is mainly responsible for larger, more automatic body movements — like those in running, walking or swimming. Its circuits, much more complex than those of the pyramidal system, contain all motor tracts between the brain and spinal cord not included in the pyramidal system. Within the brain, numerous extrapyramidal tracts link the motor cortex, basal ganglia, cerebellum, thalamus and brainstem. Along these relays, neurons form multiple

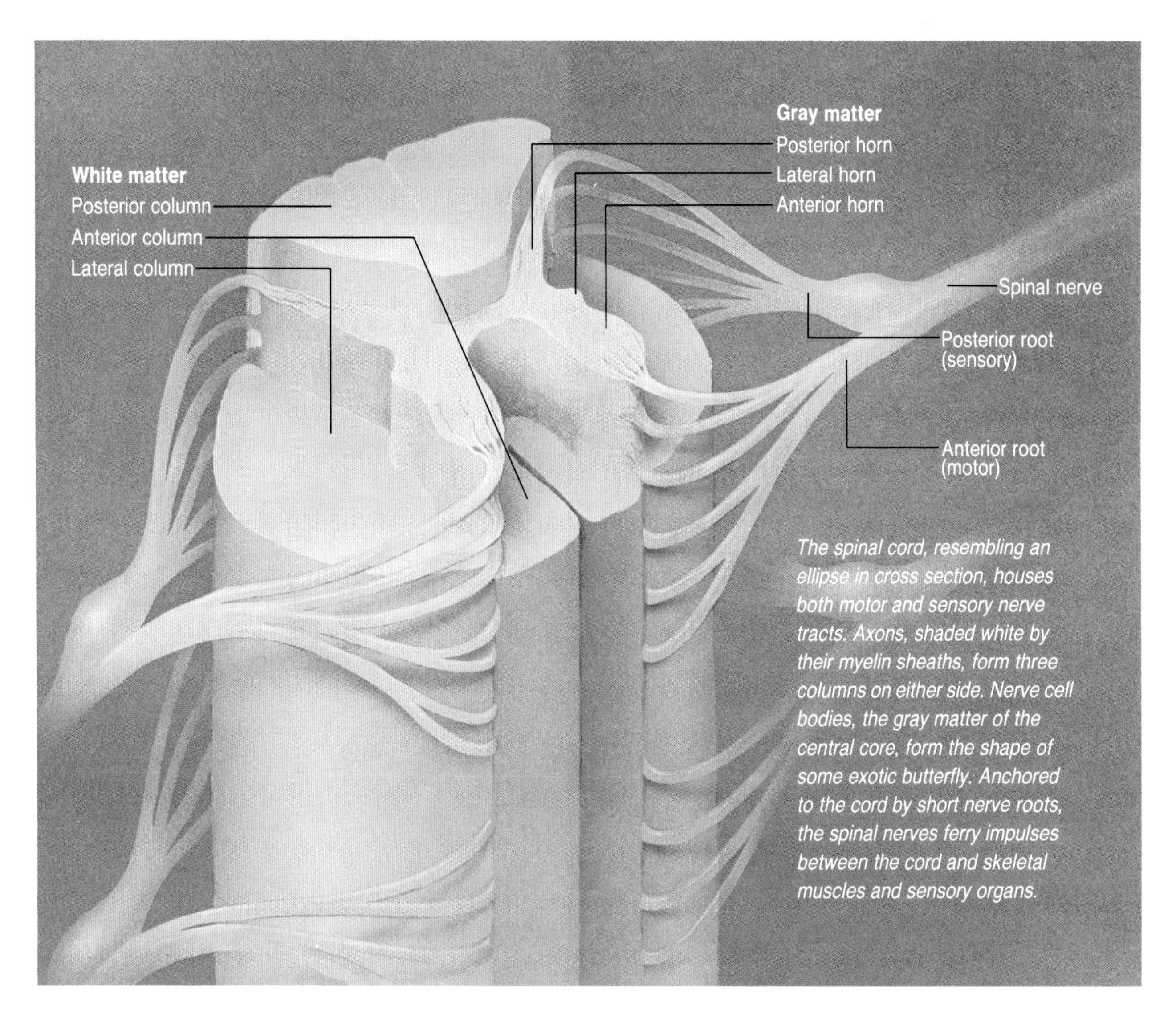

The spinal cord, resembling an ellipse in cross section, houses both motor and sensory nerve tracts. Axons, shaded white by their myelin sheaths, form three columns on either side. Nerve cell bodies, the gray matter of the central core, form the shape of some exotic butterfly. Anchored to the cord by short nerve roots, the spinal nerves ferry impulses between the cord and skeletal muscles and sensory organs.

synapses which repeatedly alter the nature of their original impulses. Some of these nerve chains descend through the spinal cord to motor neurons. Others loop back to the cerebral cortex, basal ganglia, thalamus and other brain regions, modifying transmitted impulses. While the pyramidal and extrapyramidal pathways appear to serve different purposes, they are by no means independent. One cannot function without the other. All body movements require continuous interaction between the two.

The spinal cord stands at the center of the circuits for movement. An oval, slightly tapered cylinder housed within the spinal cavity, the cord extends from the base of the cranium to the small of the back — an average length of seventeen or eighteen inches in adults. Roughly the width of the little finger, the cord is divided into nearly symmetrical halves by two deep furrows. The interior of the cord, filled with neuronal cell bodies, is referred to as gray matter. A cross section of the spinal cord reveals gray matter in the shape of a butterfly. Extending forward, like front wings, are the anterior horns, and extending backward, like hind wings, are the posterior horns. The cord's lateral horn passes through the center of the imaginary butterfly. Surrounding the gray matter are masses of axons called white matter because they are sheathed in myelin, a fatty white substance. In each half of the cord, the white matter forms three columns — anterior, posterior and lateral. Each is a large bundle of axons divided into smaller cables, the pathways that conduct sensory and motor impulses between the brain and all other parts of the body. Included among the cables are the pyramidal and extrapyramidal tracts.

Thirty-one pairs of spinal nerves connected to the cord transmit motor impulses to the muscles in the body's periphery. A pair of spinal nerves

Her spinal cord pierced by arrows, an alabaster Dying Lioness, carved in the Palace of Ashurbanipal at Nineveh nearly 3,000 years ago, still struggles courageously, despite paralyzed hind legs.

emerges at each segment of the cord, one from the left and one from the right side of the cord's gray matter. Two short roots connect each nerve to the spinal cord. The posterior root contains axons from sensory neurons housed in the posterior horn. With cell bodies outside the spinal cord, the roots of sensory neurons conduct impulses to the cord from sensory receptors in the part of the body which the nerve supplies. Composed of axons from motor neurons with cell bodies residing in the anterior horn, the anterior root conducts motor impulses from spinal cord to skeletal muscle. In most parts of the body, spinal nerves from adjacent segments overlap so that most muscles are supplied by at least two different spinal nerves. Overlap ensures that if one of the nerves is injured, the muscle, although weakened, will not become paralyzed.

If the spinal cord is severed, either through injury or through disease, the parts of the body lying below the point of severance are paralyzed, cut off from the flow of motor and sensory information. Spinal shock, an initial period which may last several weeks after the injury, causes loss of all reflex functions as well as a flaccid paralysis in the muscles below the severed region. With recovery from spinal shock, muscles of the lower limbs commonly experience extensor spasms, or straightening, and some reflex activity reappears. Should the spinal cord be severed above the fourth cervical vertebra in the neck, death from lung paralysis is almost certain. The prisoner condemned to the hangman's noose was doomed to this fate.

The neurons in the anterior horn of the spinal cord, as well as those in the brainstem supplying the cranial nerves, constitute what Sherrington referred to as the final common path. Because they are the only neurons with axons supplying skeletal muscles, all neural influences associated with movement converge on them. An average motor neuron may have as many as 15,000 synapses providing information from all parts of the body. Any condition that interferes with their ability to conduct impulses can prevent contraction of the muscles they activate.

Poliomyelitis does just that. Caused by a virus, polio crippled and killed millions until Jonas Salk developed the first vaccine in the 1950s. The disease still appears in countries lacking vaccination programs. In its mild form, symptoms are restricted to fever, headache, sore throat, nausea and vomiting, often with an accompanying stiff neck or back. The acute form of polio attacks motor neurons in the anterior horn, causing muscle atrophy and permanent deformity as fatty and fibrous tissue replace muscle fiber. Even in its serious form, however, polio does not inflict further damage once it has run its brief course.

More devastating is amyotrophic lateral sclerosis, a progressive motor neuron disorder commonly known as Lou Gehrig's disease, after the New York Yankees' first baseman who succumbed to it in 1941. Adults in the prime of life are its victims. The disease accounts for one in every one thousand deaths occurring after the age of twenty. Although the cause is still unknown, its grim course is well documented. The

The Egyptian priest Ruma, in the funeral stele below, testifies to poliomyelitis's long-standing battle with human muscle. His withered leg is the first known depiction of neurological disease.

motor nerves of the spinal cord, brainstem and cerebral cortex gradually disintegrate, and the muscle fibers they supply atrophy. The result is paralysis. In its beginning stages, muscles feel weak and unresponsive and may twitch involuntarily. When motor neurons in the brain are also affected, speech may become slurred and monotonous, and the victim has trouble swallowing and retaining saliva. Death generally follows within two or three years.

The basic building block for all voluntary movement is the motor unit, a single axon from a motor neuron in the spinal cord and all of the muscle fibers it supplies. Each muscle contains many motor units. The type of muscle determines the size of its units. Muscles that require precise control and that act rapidly have small motor units of a few muscle fibers, enabling them to increase tension gradually. In large motor units, tension increases swiftly as each additional

At the neuromuscular junction, nerve fiber and muscle fiber meet. Beadlike sacs, synaptic vesicles, release chemical transmitters across a cleft, which meanders diagonally in the micrograph below.

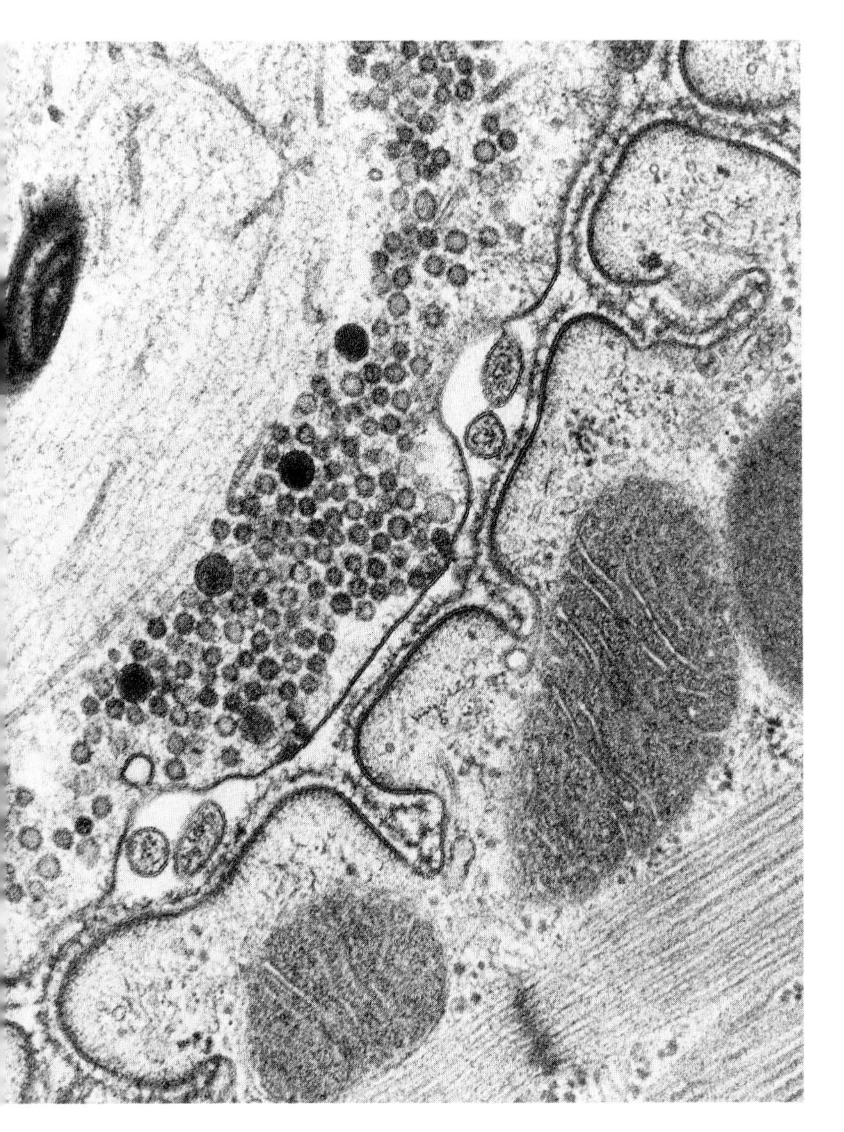

motor unit is enlisted. On average, there are about 150 muscle fibers to a unit. A single motor unit in the tiny muscles of the eye may contain only three muscle fibers, while those in the relatively slow-moving large muscles of the calf house a thousand or more. Some muscle fibers of adjacent motor units overlap so that the units contract in support of each other to some extent.

Within each motor unit, muscle fibers obey the "all or none" principle, meaning that all contract or none contracts. If the muscle fibers of a motor unit are sufficiently stimulated by nerve impulses to contract at all, they will contract maximally. Within a given muscle, however, some motor units contract while others remain at rest. Resting and active motor units frequently alternate roles, helping muscles ward off fatigue and permitting movements to be smooth.

The nervous system varies the strength of muscle contractions by precise manipulation of motor units. The same muscles that lift a twenty-pound crate are also recruited in raising a pencil. But if equal force were summoned to lift both crate and pencil, one might stab oneself in the forehead. The body tailors muscular force to the task either by varying the total number of active motor units in a given muscle or by varying the frequency with which they are activated.

The Neuromuscular Junction

Within each muscle fiber, approximately at the midpoint, lies a cleft. This is the neuromuscular junction, which receives the message to move a muscle. The electrical impulse then triggers the release of the chemical substance acetylcholine, initiating a chain of events leading ultimately to a muscle contraction. Certain diseases and toxic substances that interfere with nerve transmission at the junction can cause muscular weakness, paralysis and even death. Myasthenia gravis, the disease that claimed Aristotle Onassis, afflicts 100,000 Americans. It is apparently due to production of abnormal antibodies that impede the action of acetylcholine at the neuromuscular junction. Usually, the muscles of the eyes, eyelids, face and limbs are affected first, causing blurred or double vision, drooping eyelids, slurred speech, difficulty in chewing or swallowing and extreme fatigue of arm and leg muscles. In mild cases, rest restores muscular strength. In advanced stages, however, the disease can be crippling, rendering its victims unable to walk, talk, swallow solid food or breathe without mechanical aids. Drugs that improve neuromuscular transmission or suppress immune activity can often control the disease. In some cases, the thymus gland, lying behind the breastbone, is removed. Plasmapheresis, a new procedure which cleanses the blood of harmful antibodies, has brought remission in many severe cases.

A variety of plant, animal and chemical toxins can also impede impulses at the neuromuscular junction. Curare, derived from several species of trees in South America and used by some Indians as a poison on arrows, inhibits muscle contraction by preventing the binding of acetylcholine at the neuromuscular junction. If a large amount of curare is absorbed, the muscles that control breathing slacken and the victim dies from asphyxiation. But curare also has medicinal value. Small doses of it are given to promote muscular relaxation during certain types of surgery.

Toxins produced by the bacterium Clostridium botulinum, found in soil and animal feces, are the most deadly poisons known. Less than one ten-thousandth of a milligram can kill a man. Five hundred grams would annihilate the world's population. The botulin toxin blocks release of acetylcholine at the neuromuscular junction, causing a flaccid paralysis that makes movement impossible. Food poisoning from botulism is most often caused by improper canning or bottling. Within forty-eight hours after ingesting the bacterium, the victim experiences unsteadiness and double vision. Paralysis of speech and swallowing soon follow. Sixty-five percent of those treated still die from respiratory paralysis.

Venom from cobras, rattlesnakes, black widow spiders and certain species of fish also affect the neuromuscular junction. A delicacy called fugu, a type of blowfish found in the seas near Japan, inspired the Japanese saying, "I would like to eat fugu, but I would like to live." The leading cause of fatal food poisoning in Japan, fugu killed sixty people in 1980. Japanese fugu cooks are required to undergo special training to learn how to re-

Adding deadly insult to injury, Yanomamo tribesmen of Brazil paint their arrows with curare, made from boiled vines. Their unlucky victims will suffocate as the poison stills respiratory muscles.

Ceramist Toshiko Takaezu molds clay on a wheel. Pottery making exemplifies the interplay between movement and the senses. Visual and tactile clues guide Takaezu's hands as she shapes the pot.

move tetrodotoxin, a neurotransmission blocker. Still, there is no way to determine if all traces of the substance have been removed. If a serving contains a lethal dose, death from asphyxiation or heart attack can ensue within twenty-four hours. A nonlethal dose will generate a pleasant tingling in the fingers, toes and tongue. The Food and Drug Administration has forbidden the import of fugu to the United States.

Nerve gases, developed by the Germans in the 1930s, are swift-acting poisons that cause loss of muscle control and coordination. Their main ingredients, organophosphates, trigger the build-up of abnormal amounts of acetylcholine at neuromuscular junctions, with the result that all muscles in the body try to contract. As muscle control dissipates, breathing becomes extremely difficult and finally impossible. Because they also contain organophosphates, some pesticides can have the same lethal effect.

As the rich connections between the brain's sensory and motor regions suggest, movement is intricately dependent on sensations. To produce fluid movement, the nervous system must have access to sensory information. Sensations provide the stimulus for many movements. Trying to move a foot that has "fallen asleep" gives some idea of how difficult it is to move without sensory feedback. The nervous system collects and integrates sensory messages, and then orders an appropriate movement in response. For every movement pattern, it must arrange not only that the proper muscle contracts but also the timing of its contractions in relation to the actions of other muscles.

The special senses — sight, hearing, smell, taste and touch — provide a large portion of the sensory feedback guiding movement. But equally significant are messages sent by the muscles themselves, messages that tell us where our limbs are in space, how fast and to what extent joint angles are changing, the length and tension of each muscle and the velocity of our movements. These sensations of body movement and position are referred to as kinesthesia. The nervous system interprets kinesthetic messages and returns information to the muscles about what they should be doing and what must next be done to continue a given movement. Although most kinesthetic information remains beneath the level of conscious awareness, enough does reach the cerebral cortex to lend awareness of the relative position of each part of the body at all times. Even with eyes closed, an individual can point to the exact location of his hand, foot or elbow.

The body has other sensing devices. The tips of dendrites extending from sensory neurons constantly gather information. Sherrington classified these receptors into three basic types: exteroceptors, visceroceptors and proprioceptors. Exteroceptors, on the surface of the body, are found in the eyes, ears, nose, mouth and on the skin. Responsive to changes in the external environment, they give rise to superficial, or cutaneous, sensations. Visceroceptors, which line the walls of the blood vessels, stomach, intestines and other internal organs, respond to changes like dryness of the throat, fluctuations in blood pressure and bladder distension. Sensations of kinesthesia, as well as those of deep pressure and pain, arise from proprioceptors, located in muscles, tendons, joints and special vestibular organs in the inner ear.

Among the body's several types of proprioceptors, muscle spindles are widespread throughout skeletal muscles. These organs respond to changes in muscle length and in the velocity of muscle contractions and relaxations. Enclosed within the middle portion of a spindle's sheath and wrapped inside two types of nerve endings are several specialized muscle fibers called intrafusal fibers. Muscle spindles contract when the muscle contracts and stretch when the muscle stretches. Stretching increases the rate at which the intrafusal fibers stimulate the nerve endings they entwine, thus speeding the rate at which the nerve endings transmit impulses. Contraction, shortening of the muscle, releases tension on the

Inside a muscle fiber lies a spindle, a special sensory organ. At its center, encapsulated by a sheath, nerve endings entwined with intrafusal fibers speed signals to the brain or spinal cord when the muscle stretches. Spindles abound in muscles performing very fine movements.

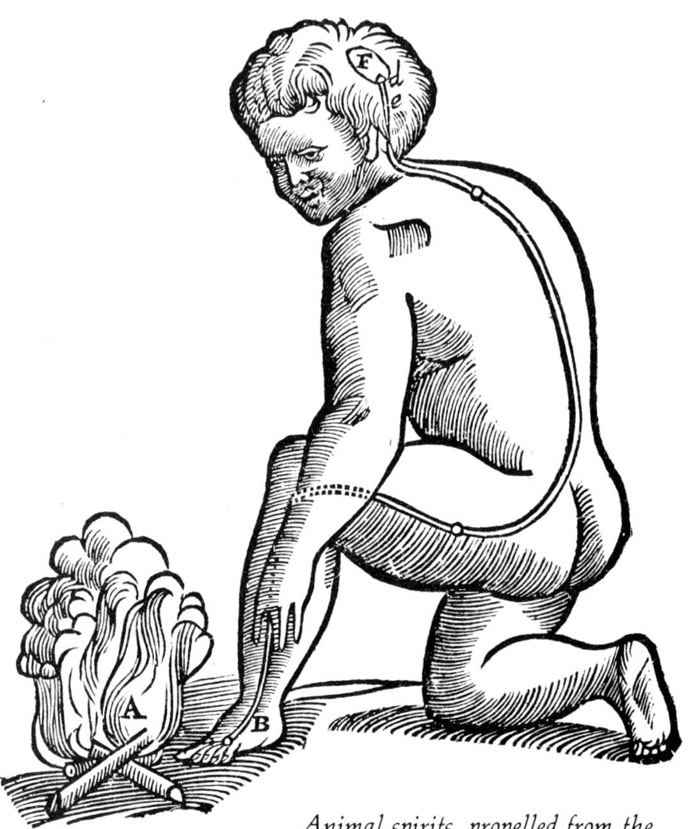

Animal spirits, propelled from the brain through tubelike nerves, generated muscle movements in Descartes's theory of reflex action pictured in his Treatise of Man. *Aroused by the fire, sensory organs in the foot pulled on "delicate threads" in the nerves, drawing animal spirits toward the muscles. Descartes mistakenly believed that contraction was the result of a muscle's inflation.*

intrafusal fibers, slowing the rate at which the nerve endings fire. Impulses traveling up to 200 miles an hour race to the spinal cord and up to the brainstem, cerebellum and cerebral cortex, permitting the nervous system to make instantaneous adjustments in the motion of muscle.

Another type of proprioceptor detects changes in muscle tension. The Golgi tendon organ is a receptor in tendons, located near their junction with muscles and linked end to end with muscle fibers. Similar to the muscle spindle in shape, the Golgi tendon organ is a capsule containing several layers of connective tissue enclosed within nerve endings. When the fibers of the attached muscle contract, they pull on the tendon, stimulating the nerve endings in the Golgi organ to fire impulses. The greater the force of contraction the more rapidly they fire. Golgi organs conduct inhibitory impulses that tend to decrease the force of muscle contraction. Some have very high thresholds of tension and respond only when muscle tension is very great, helping to prevent extremely forceful contractions that might separate the tendon from the bone. Others having lower thresholds supply the nervous system with continuous information about muscle tension.

A prime illustration of how proprioceptors guide movement is the stretch reflex, the most familiar example being the knee jerk. When the legs are crossed at the knee, and the leg above is lightly tapped at the patellar tendon just below the kneecap, both the tendon and the quadriceps muscle at the front of the thigh are stretched. Stimulated by the stretching, muscle spindles in the quadriceps transmit impulses over a nerve circuit, causing the quadriceps to contract sharply. As it contracts, the lower leg jerks upward in a kicking motion. The stretch reflex is vital to posture, which depends for balance on equal pulls of opposing muscles. If any one muscle should contract too forcefully, the stretch reflex ensures that the opposing muscle stretches and then quickly contracts, restoring balance.

Any rapid, involuntary response that is the same for a given stimulus is a reflex. Most reflexes do not reach the cerebral cortex. Instead, they function independently at the spinal cord or brainstem and, so, we are rarely aware of them.

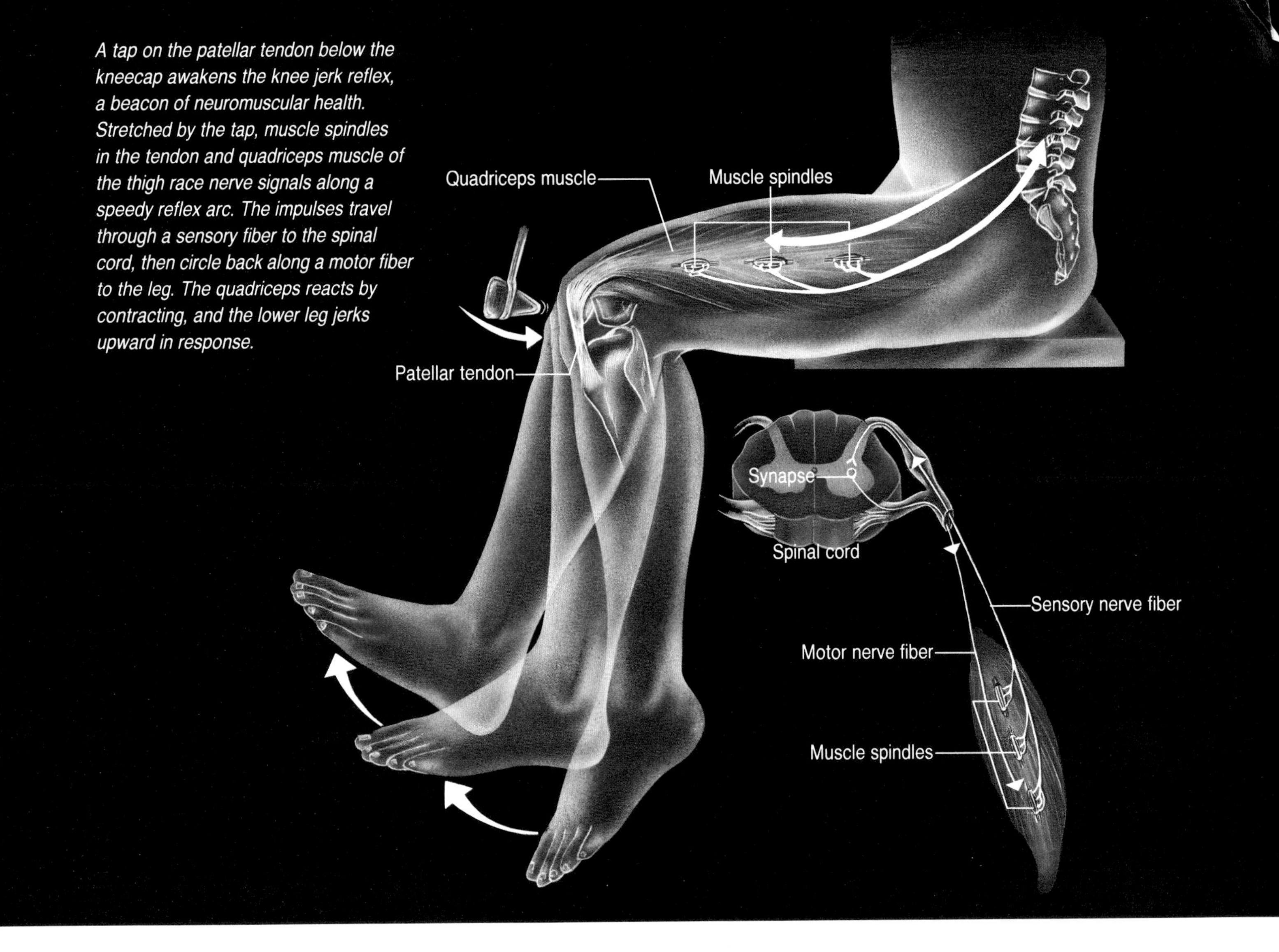

A tap on the patellar tendon below the kneecap awakens the knee jerk reflex, a beacon of neuromuscular health. Stretched by the tap, muscle spindles in the tendon and quadriceps muscle of the thigh race nerve signals along a speedy reflex arc. The impulses travel through a sensory fiber to the spinal cord, then circle back along a motor fiber to the leg. The quadriceps reacts by contracting, and the lower leg jerks upward in response.

But the human body is a veritable mass of reflexes. The mouth waters at the taste of food, the pupils constrict in response to strong light, the hand pulls away abruptly from something hot. The sneeze and the hiccup are also reflexive, as are blood pressure and heartbeat. Even during a seemingly static activity like sitting, groups of muscles continually contract and relax by reflex action to maintain balance in the body's posture.

Reflexes travel over extremely speedy circuits called reflex arcs, where they are converted from incoming sensory impulses to outgoing motor impulses. The simplest type of reflex arc is the two-neuron arc, of which the stretch reflex is an example. When muscle spindles stretch from a tap on the patellar tendon during the knee jerk test, they speed impulses along a sensory nerve fiber to a cell body in the spinal cord, where it meets a motor neuron. Traveling along a fiber from the motor neuron, the impulses circle back through a spinal nerve to the muscle originally stretched, making it contract, which results in a kicking motion. Because the two-neuron arc involves only one synaptic junction, the stretch reflex is one of the fastest known. It takes only one twenty-thousandth of a second. Most other reflexes travel over arcs involving three or more neurons. The flexion reflex responsible for withdrawal of the hand from a hot surface occurs over a three-neuron arc. Impulses from the hand travel from a sensory neuron to an interneuron, which then activates many motor neurons in the arm, hand and fingers. A painful stimulus to a tiny patch of skin may, by reflex, cause several joints to flex as the hand jerks away.

A sensory impulse accompanying the withdrawal reflex also ascends the spinal cord to the brain, where it is interpreted as pain. By this time, however, the reflex has often already been completed. Such reflexes occur whether or not pain is experienced, even in the quadriplegic whose spinal cord has been severed, or the laboratory animal whose cortex has been removed.

Because certain abnormal reflex responses can indicate a neurological disorder, doctors often test reflexes during examinations. In addition to the knee jerk, other important reflexes include the ankle jerk — an extension of the foot in response to tapping the Achilles tendon. Likewise, the abdominal reflex, the drawing in of the abdominal wall when the side of the abdomen is stroked, is a normal reflex. The Babinski reflex, which causes the great toe to extend when the outer sole of the foot is stimulated, is normally present only in infants less than eighteen months old. Beyond that age, the normal response to outer sole stimulation is called a plantar reflex, in which the toes curl under and the front of the foot is flexed. Reflex responses vary from one person to the next. Alcohol and lack of sleep can slow reflex responses. An exaggerated or diminished response can be a sign of neurological disease. More important, reflexes should be equal on both sides of the body.

The human body cannot sufficiently resist the force of gravity while standing without the support of reflex postural mechanisms. When a person stands upright, gravity causes his knees to begin to buckle. As this occurs, muscle spindles initiate the stretch reflex, causing increased contractions of the extensor muscles that straighten the knee. Equilibrium is not solely dependent on the proprioceptors. Many different types of sensory receptors continually bombard coordinating centers in the brain with postural information.

The skin, eyes and special organs in the inner ear are equally important. When walking down a dark hallway, simply touching the wall greatly enhances stability by providing the brain with messages from skin receptors. The system is so adaptable that lack of any one type of sensory input may have but scant effect on one's ability to maintain equilibrium.

Even without visual information about distance, the blind manage to keep their balance quite well under most circumstances. Damage to the vestibular organs in the inner ear, the sense organs of balance, may not seriously affect equi-

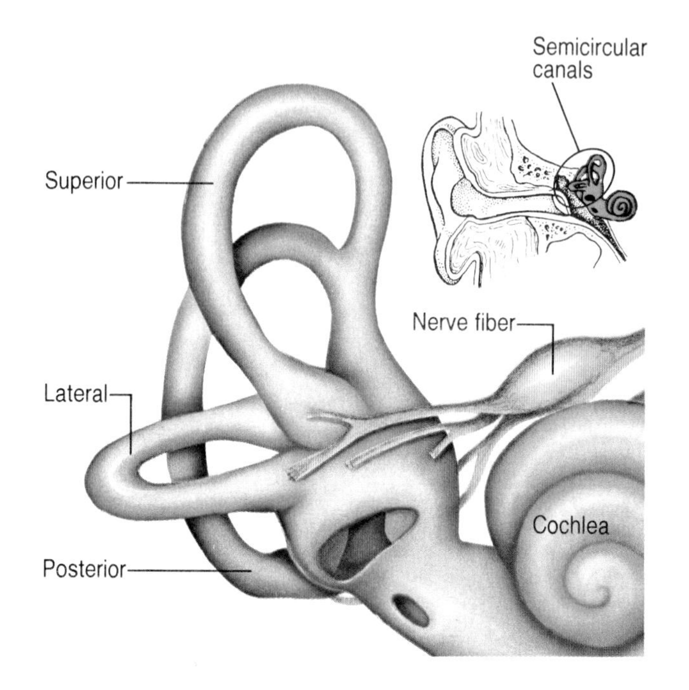

An infinitesimal tilt of the head excites the vestibular organs of the inner ear, guardians of equilibrium. Occupying the three planes of space, the semicircular canals sense changes in the body's rate of motion. Within the bony labyrinth are the utricle and saccule, membranous sacs lined with hair cells, that inform the brain of the body's position in relation to the force of gravity.

librium provided other sensory information is available. Lodged in the skull's temporal bone, the vestibular organs are three curved, fluid-filled tubes known as the semicircular canals and two large chambers, the utricle and saccule. The semicircular canals detect sensations of head movements such as angular and linear acceleration, while the utricle and saccule sense changes in the position of the head. The macula, a tiny patch on the wall of both the utricle and saccule, is covered with thousands of hair cells and a gelatinous membrane. As the head changes position, pressure on the membrane exerts a pull on the hair cells. Nerve fibers leading from the hair cells to a cranial nerve transmit continuous information to the brain about the amount of pull and, thus, the position of the head. The semicircular canals, connected to each utricle, represent the three planes of space. When the head tilts as posture changes, the fluid within the canals shifts, stimulating sensory messages about the body's position. Prolonged or rapid rotations can overstimulate the fluid, bringing on dizziness or motion sickness.

Learning to Move

The infant enters the world an appealing but clumsy creature of flailing limbs. Unable to control most of his skeletal muscles, he must begin the laborious process of learning how to move. The newborn baby possesses only a few basic movement patterns, and to these he applies unnecessarily strong effort often involving most of his body. As he begins to explore the world, his first attempts at mastering skilled movements rely heavily on sensory feedback, especially visual clues. His motions are plodding and hesitant and require great concentration. With repeated efforts, they gradually become more precise and less dependent on sensory feedback. Eventually new patterns become programmed into the nervous system, and what began as highly conscious sequences of muscle actions now proceeds automatically, each motion signaling the next. The same process takes place when learning new motor skills at any age. When a person first learns to drive a car, he must pay strict attention to every aspect of the experience, deliberately deciding how far to move the steering wheel left or right, when to step on the brakes and how much force to apply to the accelerator. Over time, the nervous system automatically produces the proper actions in sequence. If a car ahead of an experienced driver abruptly screeches to a halt, it is not necessary for the driver to think about a proper response. He immediately hits the brake.

The first two years of life mark a dramatic transformation from helpless newborn to a young child capable of reaching, grasping, assuming an erect posture and, eventually, transporting himself on two feet. Motor development follows five basic stages. In the first, an infant gains control of his upper body, lifting first his chin, then his chest as he lies on his stomach. At five months of age a baby can usually sit when supported and grasp an object. By the end of the second stage, at about eight months, he has learned to control his entire trunk, progressing from sitting in a high chair to rolling and standing with help. Active efforts at locomotion begin during the third stage, when the baby crawls on his stomach. At about twelve months, during the fourth stage, most children start crawling on all fours, and by eighteen months, have begun to walk by themselves.

The time and effort devoted to learning new patterns of movement are quickly forgotten. We take the ability to move for granted. Only an injury or a crippling disease reminds us of the difficulty of mastering ordinary daily movements. A technique called electromyographic biofeedback has proved remarkably successful in rehabilitating the physically handicapped, breathing new life into limbs that have fallen still. It is also used to relax spastic muscles and relieve tension headaches and chronic anxiety. Researchers are using the technique to learn more about muscle activity and movement patterns. The electromyograph (EMG) is an instrument that records electrical activity in muscles from electrodes placed either on the skin or in the muscle. Electronic equipment translates the information into an audiovisual display of flashing lights and buzzing sounds, providing crucial sensory feedback. With instructions to concentrate on the EMG responses, healthy patients are soon able to manipulate their muscles with striking precision. They can even make single motor units fire or not fire at will. John Basmajian, an anatomist and a leading EMG researcher, has taught healthy individuals to manipulate several motor units from different muscles in the hand simultaneously. Some subjects learn to vary the rate of firing, actually creating patterns of motor impulses that imitate subtle drum rhythms. A few can perform without any visual or acoustic cues.

Reawakening Dormant Muscles

For victims of stroke and spinal cord injury, EMG biofeedback has helped to strengthen and restore some function to muscles previously considered paralyzed. Basmajian has used biofeedback to rehabilitate 500 stroke patients, primarily by training them to use alternate nerve pathways that bypass damaged cerebral regions to control their muscles. While fine movements cannot be restored, many patients do recover some limb function. William Finley, director of the department of electrophysiological studies at Children's Medical Center in Tulsa, Oklahoma, supplements physical therapy with biofeedback for quadriplegics who have suffered spinal cord injuries. Often, such patients retain some sensation in paralyzed limbs but are not aware of it. In one study, Finley showed five quadriplegics video displays of sensory impulses in their hands, and then taught them how to increase the signals. All five patients were soon able to feel pressure and motion sensations in their hands.

More than three decades ago, researchers had begun to use EMG signals to operate artificial limbs. Over the years, advances in EMG research and the miniaturization of electronic parts have paved the way for the development of myoelectric prostheses. The Boston elbow, developed at the Liberty Mutual Research Center in Boston, restores forearm flexion to above-elbow amputees. Made of metal and plastic, the device weighs little more than two pounds and houses a battery-powered motor. The amputee can add a hook for heavy manual labor or a hand unit for light work. Electrodes attached to the skin over the forearm stump detect and amplify electrical signals generated when the amputee flexes residual muscles in his arm. By willing the hand to open or close, the signals activate the motor, and the hook or joints in the hand respond appropriately. The artificial arms of the future may permit finer movements as well as sensory feedback, which would be conveyed through microsensors in the glove tips. Researchers are also working to develop artificial feet and legs.

Even with refinement, however, prostheses driven by EMG signals will probably never permit a wide range of motion, since their dynamics are fairly cumbersome and differ in basic ways from the limbs they replace. Too, many amputees lose interest or become fatigued by the considerable concentration and effort required to move the prostheses. These problems have spurred scientists to search for alternate ways to restore movement to the handicapped.

At the Liberty Mutual Research Center, researchers are working on a neuroelectric prosthesis, a device directly controlled by nerve signals, to take advantage of what the project's director Carlo De Luca calls the "elegant programming of the central nervous system." Also the director of Harvard's Neuromuscular Research Laboratory, De Luca has successfully recorded nerve signals for movement from rabbits' severed sciatic nerves, running through the pelvis and upper hind legs — an important step in learning to record and isolate the electrical patterns within the motor nerves initiating movements. De Luca conjectures that a neuroelectric prosthesis might be available to amputees in the near future. The electrodes, implanted around severed motor nerves, would continuously record and transmit motor messages from the nervous system to a receiver outside the body, which would relay the signal to the prosthesis. The major benefits of a prosthesis controlled by nerve signals rather than muscle signals would be a greater variety of motion as well as more natural movements. To carry out a given movement, an individual would merely think about the action and the nervous system would spontaneously initiate the chain of events leading to the appropriate movement of the prosthesis.

If scientists succeed, the applications of such a technique might extend beyond prostheses for

Sir Charles Sherrington

Architect of the Nervous System

When Charles Sherrington began his career in the 1880s, neurophysiology resembled a jigsaw puzzle whose pieces lay scattered across the scientific halls of Europe and America. Theories about the structure and function of the nervous system were piecemeal at best and rife with controversy. Few scientists had attempted to correlate separate channels of research, and their fund of anatomical knowledge was meager. The solution to the puzzle of how the nervous system produced organized behavior awaited Sherrington's wealth of experimental insight and his genius for unifying diverse lines of research.

A gentle, modest man who wrote poetry in his later years, Sherrington was a scholar of art, history, philosophy and architecture as well as science. His broad interests were shaped largely by his step-father, Caleb Rose, an archeologist and classical scholar. Although the Rose home in Ipswich, England was a rendezvous for artists, Rose, perhaps awed by the boy's phenomenal memory and acute powers of observation, steered him toward biology. After undergraduate work at Cambridge University, Sherrington spent three years in France and Germany, study-

ing with some of Europe's leading scientists. At first drawn to pathology, he decided upon his return to England in 1887 to concentrate instead on neurophysiology.

Faced with the intimidating complexity of the nervous system, Sherrington sought first to reduce it to what he saw as its basic functional unit — the reflex. To build the anatomical foundation necessary for physiological study, he spent the next decade at what he later described as "boring" and "pedestrian" research — tracing the distribution of motor and sensory nerve fibers in the spinal cord.

While immersed in anatomy, Sherrington began to formulate a comprehensive view of the motor functions of the spinal cord. From these studies emerged the principle of the reciprocal innervation of antagonistic muscles, describing the nervous control of opposing muscles. This theory, proclaimed Sherrington's contemporary, physiologist Edgar Adrian, was "the clue to the whole system of traffic control in the spinal cord."

With the publication in 1906 of *The Integrative Action of the Nervous System,* compiled from lectures, Sherrington was hailed as "the main architect of the nervous system." A watershed achievement synthesizing the work of an entire era and opening the way to a new one, the book presented the first comprehensive, experimentally documented explanation of how the nervous system produces coordinated movements.

Sherrington collected many honors during a career that spanned more than fifty years. For his research on the motor unit, the basic unit of movement, he was awarded the Nobel prize in 1932. When he died in 1952 at the age of ninety-four, colleagues mourned the passing of the "supreme philosopher of the nervous system."

A prototype prosthesis designed by Japanese researchers, bottom, waves to the future and its robot predecessor, top. Incorporating sensors and a microprocessor, the unit would restore hand functions to amputees.

Quadriplegic Greg Kramer pedals a stationary bicycle with the help of electrical stimulation and computerized feedback. Jerrold Petrofsky, who developed the system, monitors the multipanel computer.

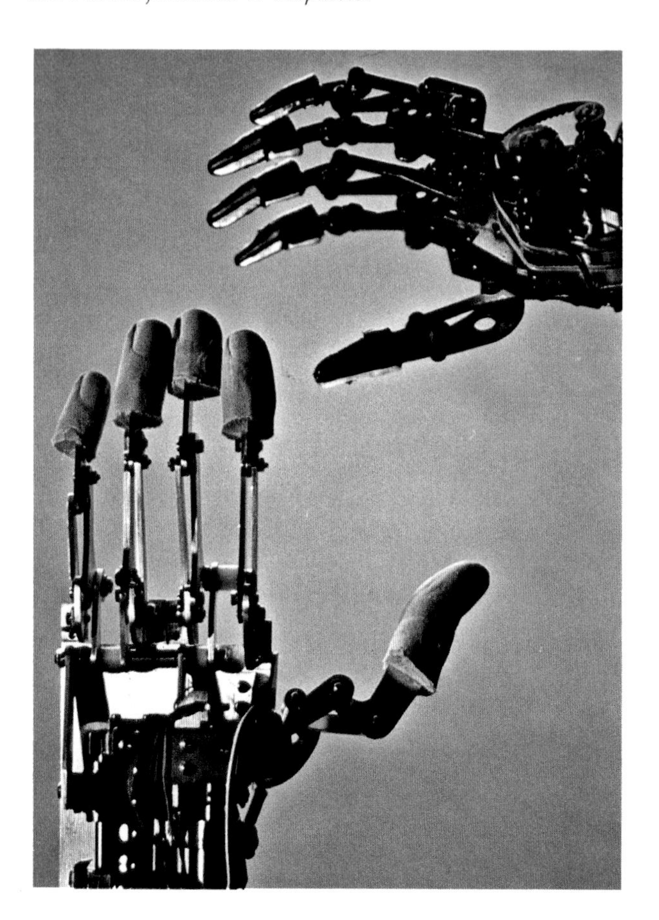

the handicapped. "Although numerous scientific and technical obstacles remain," De Luca thinks, "it is conceivable that in the future it will be possible to activate and control electromechanical devices external to our body via a direct link from the nervous system. The possibilities are really endless. Everything which is now electrically operated could theoretically be controlled by nerve signals."

An equally exciting approach to the artificial stimulation of natural muscle movements is the brainchild of Wright State University's Jerrold Petrofsky, a professor of engineering and physiology. For the last thirteen years, Petrofsky and his colleagues have been attempting to program a computer to imitate the elaborate motor operations of the nervous system. People with extensive nerve and muscle damage due to muscular dystrophy or multiple sclerosis cannot be helped by electrical stimulation, but those who are paralyzed from spinal cord injury or stroke might regain some movement from Petrofsky's technique.

The computer, a tiny microprocessor installed above a paralyzed limb, would induce movement by transmitting electrical impulses through electrodes implanted in the motor nerves of the paralyzed limb. Skin sensors attached to the microprocessor would detect electrical activity in muscles normally used in walking, and would alert the computer to contract leg muscles when the person wants to walk. In the spring of 1982, Petrofsky's microprocessor enabled a quadriplegic to pedal a stationary bicycle for five minutes. The patient's movements, observed Petrofsky, "were perfectly natural." In this preliminary trial, the computer was not implanted. Instead, multiple electrodes were taped to the skin of the legs, feeding information to the computer, which then sent impulses back to the muscles to make them contract. To sustain longer periods of activity, Petrofsky's subjects must undergo training to strengthen atrophied muscles.

Petrofsky's undertaking is enormous. He must devise a system that would control nerve stimulation with painstaking precision to ensure smooth muscle contraction. To do this, he must analyze nerve and muscle patterns, and translate them into a computer program based on mathematical models that take into account the myriad factors influencing muscle performance. Perhaps most awesome is the task of smoothly regulating the coordinated movements of groups of muscles. Despite these obstacles, he believes the microprocessor will provide a new form of exercise therapy that would help reverse the bone and muscle deterioration of paralysis, and ultimately restore movement to many paralyzed people.

Petrofsky's microprocessor and De Luca's prostheses would permit a more normal life for thousands confined to wheelchairs and other artificial aids. To Petrofsky's optimistic predictions, however, De Luca adds a note of caution. "If you think in terms of what we are doing, one can be rather euphoric. I would dampen that enthusiasm somewhat. . . . When it comes to joining machines with the human body, there's an awful lot of physiological information that we need and simply do not have."

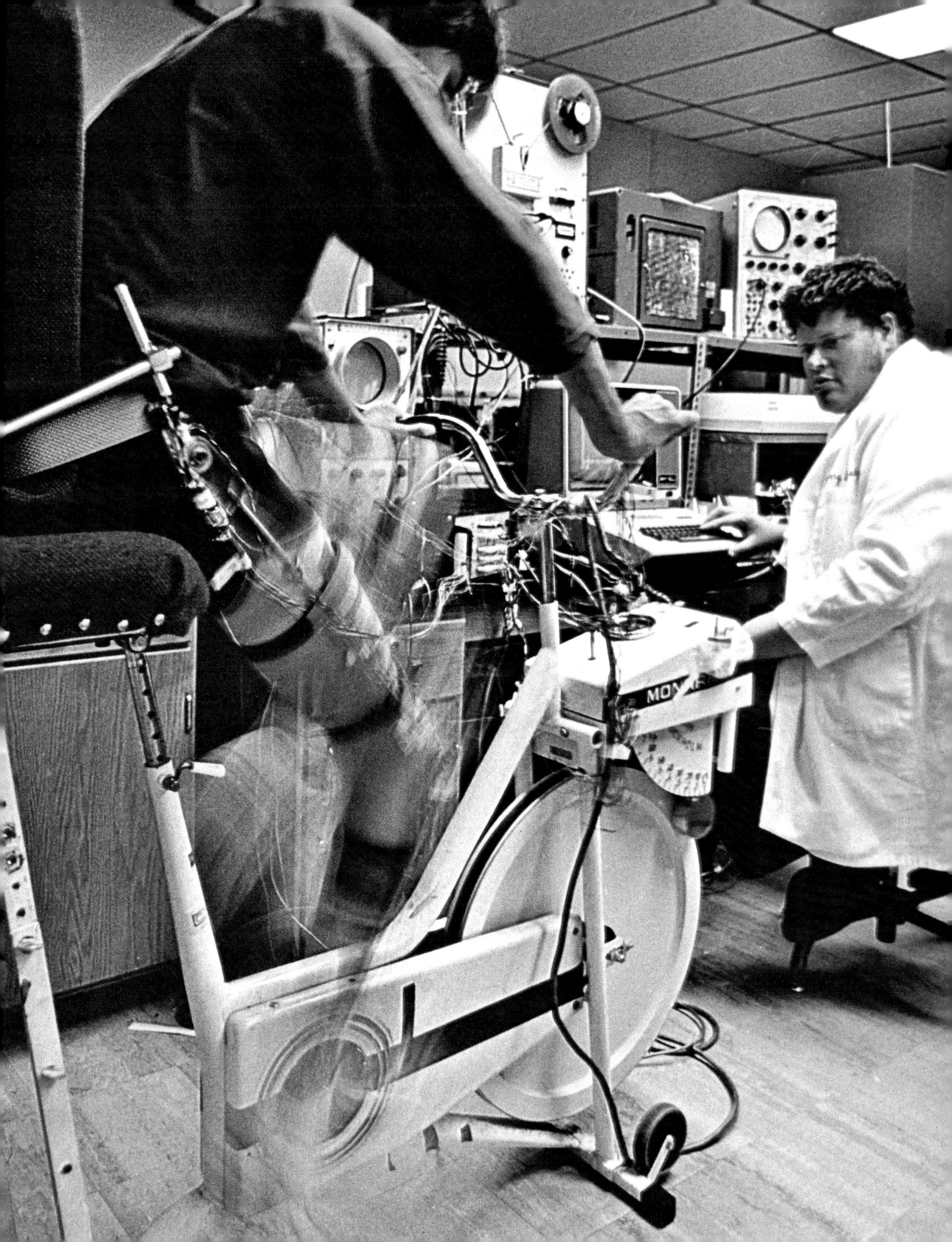

Chapter 5

The Language of Movement

Muscles fulfill a dual purpose. They give us movement, which enables us to survive, but they also lend expression to the mind. Many slight motions are not perceived because the muscles can act without conscious direction. Others, however, are the result of muscle contractions willed by the mind. Thoughts and feelings resonate through the body and become action. Only through muscles are emotions released and shared.

When we look at our hands, we sometimes marvel at their dexterity and resilience, their ability to adapt to any task of precision or gripping tension. Scarcely ever do we consider our feet in this light because they are not so adept as the hands, serving, as they do, to support and propel us. But in the evolutionary history of mankind, the development of the foot is more important than it may first appear. All primates can grasp objects, sometimes with their feet and tails as well as their hands, but most move in an awkward, tottering gait when they must walk on land. Man's upright posture meant that his hands were no longer needed for locomotion; they were free to gather food and design tools.

Within man himself, unseen organs pulse with the steadiness of a timepiece. Patterns of movement unite all of Nature in a rhythmic timetable: the winds, the change of seasons, the flow of rivers and tides, the migratory journeys of fauna and fowl. With time, early man learned which movements were useful and which availed him little. He also learned that he possessed grace and could move with a pleasing sense of ease and excitement. He joined with others in a shared language of movement that had meaning beyond the chores of everyday life. Together, they had created dance.

Because children can dance for joy before they learn to speak, some people think dance may have been the first form of communication. Body

Muscles are the tools that the mind and spirit use for expression. Without muscles, the chanteuse in Edgar Degas's painting of 1878 would be mute, her song unaccented by gesture. Movement becomes its own language, whether understood in the slightest communication or in the grand traditions of dance.

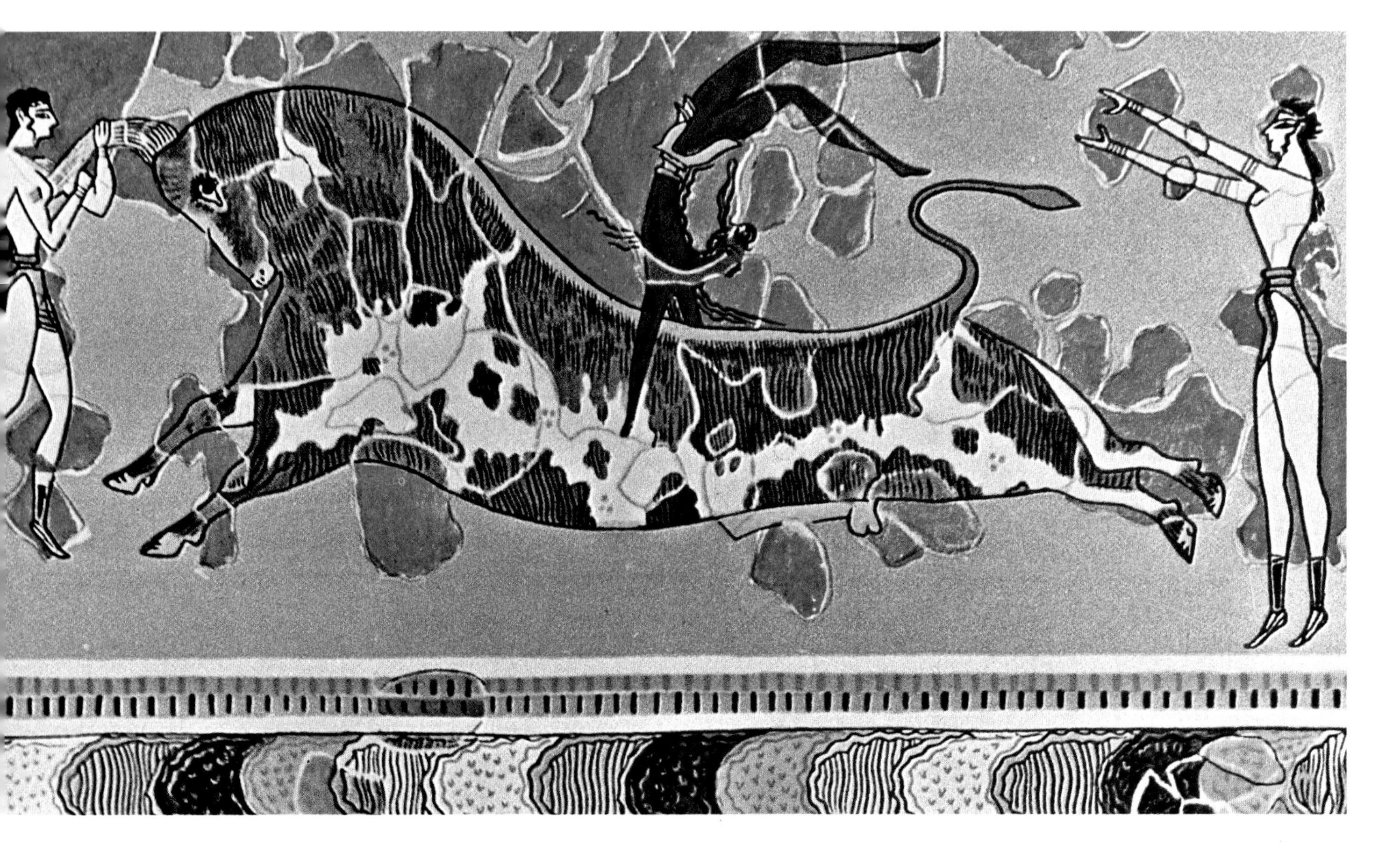

movements reflect a culture's mores, so the first dances may have created a shared bond before there were words to describe what that bond meant. Ancient myths claim that the earliest dances imitated the movements of animals. Thousands of years ago, in caves in France and Spain, primitive artisans painted lifelike images of animals in motion, exhibiting an appreciation for the fluidity of movement. In an animistic world in which animals, trees and the cycle of seasons were invested with spiritual power, it was natural to think of these objects or events to mark memorable moments in life.

Dance was a binding force in early societies, lending a common identity to a group of people. To celebrate birth and death, puberty and marriage, the success of hunts and harvests, men and women joined in dance. Together they could invoke a common deity and preserve traditional customs. Dancing intensified religious experience by uniting physical with spiritual energy. The more vigorous the movement, the more pleased were the divine powers. Unlike modern dance, primitive dance was not performed for aesthetic pleasure. Even today, the tribal dances of isolated peoples, though colorful and exciting to behold, serve primarily religious purposes. Throughout Africa, dancing continues as an integral part of community life. Body painting and elaborate headdresses often help evoke the spirits of animals and gods; drum rhythms determine the steps and movements of the dancers. Similar practices of body painting and rhythmic intensity prevailed in the dances of the American Indians. Often strenuous tests of physical endurance, many of their dances honored the religious importance of animals and birds. Reenacting tribal myths, bolstering courage before battle or promoting the fertility of the fields, dance retained its spiritual mystique.

On the Mediterranean island of Crete, 1,500 years before the birth of Christ, acrobatic dancers leaped across the backs of charging bulls, opposite. The spiritual importance of these daring maneuvers is not known, but in East Africa, dance is still a significant part of the religious life of the Watusi tribe, above. In the remote Himalayan kingdom of Bhutan, left, royal dancers don colorful costumes and masks representing deities and other symbolic figures. The skull mask and red-nailed gloves worn by the dancer at left signify death.

In Bali, near Indonesia, dance is a part of everyday life. By his wide-eyed face of terror, a performer in the religious barong *dance admits his fear of the unknown and allays evil influences that may beset him.*

Frequently called the oldest art, dance flourished in ancient Egypt. Relief carvings and wall paintings executed over four thousand years ago depict the vigorous, athletic style of Egyptian dance. Leaps, kicks and backward flips were common. With origins in religious celebrations, many of these acrobatic displays were led by priests. "They made a thousand jests with their feet and heads," wrote Roman author Apuleius. "They would bend down their necks and spin round so that their hair flew out in a circle."

From Osiris to the Orient

Most Egyptian gods were honored through ritual dancing, but none was so revered as Osiris. His death and resurrection came to symbolize the cycle of the Nile, the river flooding and renewing parched soil. Dancing also became an international spectacle in Egypt, and its patterns grew progressively complex. Early written records from Egypt tell of traveling professional troupes that joined in ceremonies to please the nobility. Pygmy dancers from the African interior appear to have been especially popular. Hindu women from Asia introduced graceful, sinuous movements. The belly dance and other erotic dances are said to have begun in Egypt.

In ancient Greece, dance became part of formal education. Socrates, Plato and Aristotle agreed that no education was complete without dance. According to Socrates, the best dancer would make the best warrior. Youths mimed sport in dance, mostly as a means to honor the gods of Greek legend. The most exuberant dances were performed by the bacchantes, the female followers of Dionysus, god of wine. The god's presence was summoned through intoxication, dances of whirling frenzy and races across the countryside. Men then performed the dithyramb, an impassioned ritual of song and dance that lauded the virtues of wine and fertility. With time, the unrestrained Dionysiac rites were refined. Traditional legends were featured in chanted hymns. Dance, poetry and song came together in one art form and gave rise to Greek tragedy. Comedy grew from bawdy theatrical traditions and became associated with the kordax, a lewd, masked dance that later evolved into slapstick comedy.

Scarf dances of modern China are heirs to a long tradition. More than two thousand years ago, a poem told of dancers who waved their long, flowing sleeves in graceful patterns "like crossed bamboo stems."

Before they were overrun by the Romans, the Etruscans had a flourishing civilization in central Italy between the seventh and fifth centuries B.C. Frescoes attest to the popularity of dance among the Etruscans. Some of the movements portrayed resemble the steps of today's modern dance. In Rome itself, dance lost the esteem it held with the Greeks. Dancing was a sign of questionable morality. In the first century B.C., Roman orator Cicero ventured the opinion that no man danced unless he was insane. Pantomime and drama replaced dance as popular forms of entertainment.

As the tradition of dance languished in Europe through the Middle Ages, it continued in the Orient according to long-held religious customs. To this day, much of Oriental dance retains a spiritual significance that is not as apparent in Western dance. In the Asian tradition, the stage itself is a symbol of the world where one must struggle to please the divine powers. The techniques of much Oriental dance are so subtle that they easily elude Western audiences. Instead of executing gravity-defying athletic leaps, Asian dancers are more likely to keep their feet on the ground and convey meaning through slight and studied gestures of the hands, arms and upper body. Movements are slow and controlled, complemented by elaborate costumes and masks.

Asian dancers begin training at an early age to learn the double-jointed movements that look physically impossible to outsiders. Each move also carries a symbolic meaning, an understanding of which is essential to mastering the art form. In just one style of dance from India, for instance, there are thirteen head gestures, thirty-six glances, nine movements of the eyelids, six each of the nose, cheeks and lower lip, seven of the chin, nine of the neck, and sixty-seven hand gestures. The upper body, legs and feet have a comparable vocabulary.

Enacting a tale of moral intrigue, a Kabuki performer in ornate costume and make-up strikes a pose. A tradition in Japan for 300 years, Kabuki is marked by slow movements laden with symbolic meaning.

Much formal dance in Japan relies on traditions alien to the West. Here, the origin of dance is related to the creation of the earth and the spreading of light. Before some dances, the performers follow rituals of fasting and purification of the body. Complex feelings are expressed through smooth movements, pauses and the overall effect of an entire performance. The gestures and gliding steps of classic Japanese dance are not nearly so physically demanding as the flying leaps and lifts of Mikhail Baryshnikov or Rudolph Nureyev. Psychological intensity and simplicity of movement are a Japanese dancer's goals. Dancers become more admired with age; they are sometimes dubbed "national treasures," given living allowances by the state, and they perform well into their sixties.

A Heritage Common and Cultured

There are two main styles of Japanese dance. One, which includes *No* drama, is derived from religious ritual and court entertainment; the more popular form, Kabuki drama, evolved from folk culture. The term Kabuki is drawn from the three language symbols meaning "song," "dance" and "skill." The first Kabuki performances, around the year 1600, were led by Okuni, a former priestess who mocked Buddhism. The original performers were women rumored to be prostitutes. Women were soon proscribed from performing Kabuki, and a tradition of male performers has endured for 300 years.

Accompanied by simple folk music, Kabuki actors perform stylized choreography before painted scenery. Costume changes usually take place in full view of the audience and are metaphors of transformation woven into story. A complicated plot, built of episodes, mounts to a dramatic climax. Conflicts of morality and passion are standard, but by imprinting their own style on the unchanging conventions Kabuki masters can give added psychological depth to often repeated plays.

Throughout the rest of the Orient there remain sophisticated dances traditionally practiced at royal courts before there was any contact with the West. The expressive gestures and movements that look merely beautiful to the untrained Western eye actually carry fine shades of meaning as events from the religious mythology of the East are retold through dance.

Folk dancing in European society has almost vanished, but it once held nearly as prominent a place in everyday culture as in the Orient. The origins of folk dance lie in the simplest of impulses. It celebrated a humble people's responses and entreaties to natural and divine forces. The folk dances that survive, if only in ceremonial form or in the pages of literature, bespeak an earlier belief in magic, fertility rites and worship of nature. In maypole dances, for example, the maypole itself represents a tree, fruitful and protective, rising from the earth as an emblem of fertility. Ribbons connecting the dancers with the central pole symbolize vestigial branches of the tree. To celebrate weddings, the passing of seasons or the sowing of crops, peasants gathered to give action to their emotions.

Both peasants and gentry developed social dances, which helped cement the bonds of the community. By the time the Renaissance dawned in the fourteenth century, dance had become one of the most popular social activities in Europe. For courtiers, rules on the propriety and execution of dances were promulgated. From southern France, the couple dance emerged, in which men and women held hands and paired off. The dances of the upper classes were more elegant and worldly than the unpolished rustic dances.

Boisterous dances, enlivened by peasant influences, became popular in England in the sixteenth century. Queen Elizabeth I enjoyed dancing and was particularly adept at the galliard and volta. The Spanish ambassador was shocked when the spinster queen appeared with skirts shorter than the fashion and danced the volta, in which couples embraced and leaped in the air.

In France during the seventeenth century, a dance was born that exemplified the classical artistic tastes of the time. The steps of the minuet were calculated with the scientific precision of baroque architecture. Refined from folk dances, it followed music with three beats to the measure. The minuet later became the essential third movement in the symphonic form developed by Franz Joseph Haydn and C. P. E. Bach.

In the seventeenth and eighteenth centuries, the minuet was the most popular dance among Europe's upper classes. Beneath musical notation, dance master Kellom Tomlinson traced the minuet's delicate steps.

Delicate steps, carefully prescribed positions of the feet and angles at which the body was held characterized the minuet. Failing to master the intricate footwork of the minuet could make a clumsy young gentleman a social outcast. Shortly after the minuet gained its pre-eminence, French dramatist Molière wrote with deliberate exaggeration: "There is nothing so necessary to men as the dance. . . . All men's misfortunes, and the appalling disasters of history, the blunders of statesmen and the errors of great generals, they have all occurred for lack of knowledge of dancing." For more than a century, the minuet remained the most popular social dance in Europe: many believe it may well have been the first dance craze in history.

Near the end of the eighteenth century, as a wave of egalitarian spirit swept through Europe, carrying the French Revolution on its crest, the aristocratic minuet began to lose favor. But the public demand for dancing persisted. Paris alone had 684 public ballrooms in 1789. Soon a vibrant new dance from Germany and Austria dethroned the minuet. With lively turns to a bouncy rhythm, the waltz became Europe's paramount dance by the early 1800s. It was easier to learn than the minuet, and people frequently devised variations on the basic steps. Its intimate physical contact, while drawing moral censure, did nothing to impede its popularity. Under the batons of the Strausses of Vienna, the waltz conquered the concert hall as well. Throughout Europe in the nineteenth century, nobleman and commoner alike danced the waltz.

To Interest the Heart

Since the turn of the century, a number of dances from the Americas have supplanted the waltz's popularity. Jazz, with its syncopated rhythms and unexpected harmonies, emerged from the black enclaves of the United States. It was a new kind of music and brought new animated dances with it. During World War I, the fox trot, so named because of its rapid, running steps, came into vogue. Just before World War II, the jitterbug became the rage. This dance required the male partner to lift the woman off the floor in truly athletic lifts and turns. Since the 1950s, young people the world over have twisted, jerked, frugged, boogalooed, swum, Watusied, hustled, boogied and jived to the insistent beats of Motown, rock 'n' roll and the programmed pulse of disco. With the frenetically paced punk rock of the late 1970s, dancers hurled themselves into one another like human bumper cars.

Social dancing of the Renaissance eventually led to ballet, the grandest tradition of Western dance. As the Middle Ages drew to a close, increasing wealth made the upper classes of Europe more interested in cultivating social pleasures. Itinerant jugglers and troubadors became the first dance masters, often instructing the nobility in posture and etiquette. The first such master known by name was Domenico da Piacenza, who published a manual of dance in 1416. Domenico described various popular dances, including several original ones of his own. By the 1500s, dance and music were common public spectacles.

With expansive murals, Thomas Hart Benton captured the variety of American life in the 1920s and early 1930s. In a typical dance-hall scene, below, couples awkwardly embrace as they step with the music.

When Catherine de Médicis became queen of France in 1547, she brought the dance with her from Italy. By the time Louis XIV came to the throne a century later, grand entertainments were frequently staged at court. Plumed horses, lavish fireworks and sumptuous decorations were the grandiose precursors of modern ballet. Louis himself danced in his youth. He became known as "The Sun King" after dancing the role of the sun in the 1653 production of *Ballet de la Nuit*. In 1661, Louis XIV established the French royal academy of dance to oversee the performance of ballets at court.

As ballet advanced toward the eighteenth century, professional dancers became ever more popular. Marie Camargo, a French dancer of the first half of the eighteenth century, was the first to shorten her skirt above the ankles and perform in the flat, thin-soled slippers that every dancer wears today. Marie Sallé, Camargo's rival, emphasized the dramatic side of ballet by unifying music, costume and dance. The high, athletic leap practiced by men, the *grand jeté,* was perfected by Florentine dancer Gaetano Vestris in the second half of the century. Drama became more important in ballet; standard pantomime gestures came to represent various emotional states. A French ballet master of the eighteenth century, Jean Dauberval, said pantomime "is a universal language, common to all time, and better than words it expresses extreme sorrow and extreme joy." Affirming movement's expressive power, he added, "I do not want just to please the eyes, I must interest the heart."

The nineteenth century gave birth to the prima ballerina, the featured leading lady of a ballet. In the first half of the century, Marie Taglioni was the favorite of Europe because of her graceful style and lithe, distant beauty. She is believed to have been the first to popularize the *en pointe* technique of balancing her weight on her toes. Now a standard practice, toe dancing requires great strength in the leg and ankle.

Once the formal style and repertoire of ballet were established in the 1800s, innovators in our own century began to challenge the accepted conventions. Between 1909 and 1929, the Ballet Russe, a Russian troupe performing primarily in

With a dual exposure, the camera catches ballerina Lisa Bradley, left, in a flourish of poetry and strength. From a standing position, she soars into an arabesque, her body poised delicately en pointe.

Paris, changed perceptions of dance permanently. It was a time of great social and artistic ferment. Ballet Russe's impresario, Sergei Diaghilev, brought together choreographer Michel Fokine, composer Igor Stravinsky and artists such as Marc Chagall and Pablo Picasso to design sets and costumes. With works such as *Petrushka* and *The Firebird,* ballet would never be the same again. Stravinsky's music was dazzling and jarring; the stage designs brash; the choreography scandalously bold. Diaghilev's leading dancer, Vaslav Nijinsky, captivated and enchanted audiences with an exotic mysteriousness.

In 1912, Nijinsky began to choreograph his own ballets. With angular, flattened movements in his version of Claude Debussy's *Prélude à l'après-midi d'un faune,* performers danced entirely in profile, creating a flattened effect reminiscent of ancient Etruscan tomb paintings. Nijinsky portrayed the mythological, half-human faun

From humble pleasure in movement, dance has evolved into the body's supreme vehicle of artistic expression. Aloft, almost like creatures born to the air, these dancers blend in a vision of grace, exemplifying the question asked by poet William Butler Yeats, "How can we know the dancer from the dance?"

Garlanded in petals of silk, below, legendary dancer Vaslav Nijinsky whirled like a flower carried on the wind in the 1911 ballet Le Spectre de la Rose. *His later years were plagued with insanity and tragedy.*

with an unmistakable sensuality. The following year, Nijinsky choreographed Stravinsky's new composition, *Le Sacre du printemps* (The Rite of Spring), creating what playwright Jean Cocteau called "a pastorale of the prehistoric world." The music's dissonance, savage rhythms and explosive orchestrations celebrated the primitive ritual of rebirth, spring breaking winter's grip on the land. Nijinsky's choreography was no less revolutionary than the music. The dancers appeared in baggy costumes based on Russian peasant dress. In defiance of traditional ballet, they moved with feet pointed at awkward angles and hands outstretched and rigid. The opening performance so upset the fashionable Parisian crowd that it rioted, and the violence carried out into the streets. It was as if the ballet awakened in the audience primal impulses long lulled by the veneer of civilization. Nijinsky had taken dance back to its dark beginnings.

Before the Ballet Russe had scored its dramatic successes in Paris, Isadora Duncan, a young American dancer, had impressed audiences worldwide with ballets that were more expressions of pure movement than the interplay of calculated steps and gestures. Inspired by the ideals of Greek art, she danced in loosely flowing gowns that created a textural complement to her fluid dance steps. By 1930 another American dancer, Martha Graham, capitalized on the importance of costume to create stunning visual and expressive effects. She intensified simple acts such as breathing into dramatic presentations of physical energy. Merce Cunningham, a choreographer since 1942, abandoned the vocabulary of traditional dance to emphasize the excitement of pure movement. His dancers were allowed to move freely, without following a strict pattern of steps. His work began what is known today as the postmodern school.

Admired for its delicate beauty, ballet is unquestionably the most arduous of the major art forms. Ballet dancers begin their training at an early age. With repeated exercise, their joints and muscles become exceptionally flexible. The stretching exercises of dancers are so effective at loosening tense muscles that they have been adopted by collegiate and professional athletic teams. Edward Villella, a former principal dancer with the New York City Ballet and an amateur boxing champion, once said, "It takes more strength to get through a six-minute *pas de deux* than to get through four rounds of boxing."

In spite of its long history and importance in our culture, dance still remains an ephemeral art, disappearing as quickly as it is performed. The music that accompanies dance can be played over and over because it is recorded and printed in notation that every musician understands. But complicated choreography is seldom recorded in a systematic way. Dance masters have attempted to develop methods of notation since the fifteenth century. Occasionally their schemes have proved priceless in preserving ballets that would otherwise have been lost. The personal notebooks of Nikolai Sergeyev kept the repertoire of the Russian Imperial Ballet alive when dancers fled the Soviet Union after the 1917 revolution.

Rudolf Laban

The Artist of Movement

As a youth, Rudolf Laban may have unwittingly caught a glimpse of his future when, one morning in 1894, he witnessed an especially beautiful sunrise. Alone in the mountains, he wondered whether he could capture the beauty of the moment in music or in paint. But "it was all too rich for that," he later recalled. He could think of only one way to express his joy. "When my body and soul move together they create a rhythm of movement; and so I danced."

Laban went on to study art and architecture, but, perhaps inevitably, he turned back to that early dawn of enlightenment. He became a dancer, a choreographer, a teacher and one of the founders of modern dance in Europe.

A philosophical man, Laban believed that all of nature moved in purposeful rhythm and harmony. Man himself was part of this universal dance. "In his gestures," Laban said, "man changes the positions of his body and limbs in space exactly as . . . the electrons, atoms and molecules of matter do." Unifying mind and body, movement could lift man spiritually. To Laban, art was present in daily experience, and dance showed that the "the aim of man was his festive existence."

From North Africa to North America, Laban sought the roots of festive life in other cultures. In ancient texts from Asia and Mesopotamia he discovered what he believed to be the archetypes of dance notation symbols. He knew that many modern styles of dance notation were limited because they could show only one movement at a time. He saw the need for a system that could record simultaneous movements. In Labanotation, the method he devised, variations in the length, shape and shading of rectangles indicate a movement's rhythm, direction and height from the floor. Dots, lines and slashes, arranged in vertical columns, denote gestures of the arms, head and hands. Labanotation can record any human movement and has enabled modern dance companies to revive early works by George Balanchine and other masters.

Laban was known throughout Europe for his innovative choreography and his schools of modern dance. He rose to the position of director of movement for the Berlin State Opera, but his theories of free individual expression were declared "against the state" and banned in Nazi Germany. In 1938, he arrived in England penniless and gaunt from hunger. To earn money, he began to teach and quickly gained prominence. During World War II, he studied the movements of factory workers and designed exercises to promote efficiency. He also advocated physical education and dance for schoolchildren.

Until his death in 1958 at age seventy-eight, Laban lived in England, yet he remained something of an enigma in spite of his accomplishments. His education, his travels, even his full name were cloaked in mystery. Over six feet tall, he exuded an inner calm and charismatic power that held his students in awe. Women found him magnetically attractive. Enormously influential in his own time, Laban seems to have been eclipsed by the passing years. Still, his notation survives, giving voice to the silent art of movement.

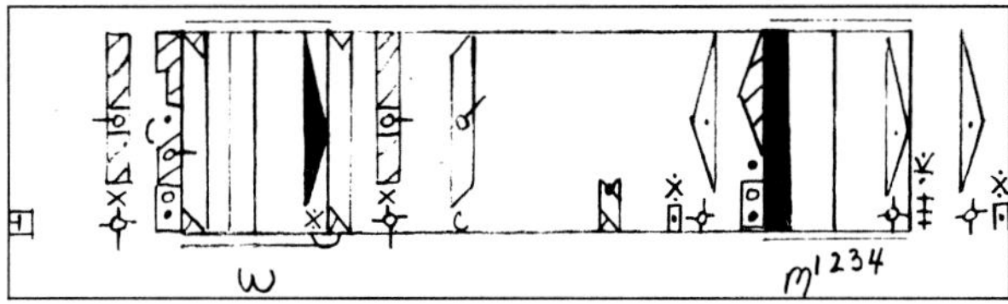

For centuries, dance masters have sought ways to maintain precise records of movement that would allow performances to be faithfully reproduced. The most widely used method of dance notation today was devised in the 1920s by Rudolf Laban. Skilled practitioners of Labanotation, as his method is known, could recognize the movements of the dancers at top in the diagram above. The cryptic-looking notation registers every exuberant kick, each outstretched arm.

Most systems of notation have proved inadequate. In the 1920s, Hungarian-born dancer and movement theorist Rudolf Laban devised a system using marks resembling musical symbols to indicate a performer's every move during a dance. Labanotation is the most widely used system today. Another technique, choreology, was later developed. Both methods, however, are time-consuming. At best, it takes about six hours to transcribe one minute of dance onto paper. Another shortcoming of dance notation is that few people understand it. Only 1 percent of dance professionals can interpret any form of notation. Help may be coming, however, from computers that present a visual image of movements on a screen. Once stored, such records alleviate the need for arcane diagrams of notation.

Speech Made Visible

Dance can be defined as movement that is stylized and ordered in such a way as to give it expressive meaning. But to some extent, all bodily movements are expressive and transmit information from one person to another. In recent years, students of body language have analyzed gestures and other movements and concluded that our actions often tell more about our characters than words do. People from many fields study the psychology of movement to learn how body language can affect their professional lives. Much modern advertising relies on the unspoken appeals of movement, gesture and posture.

A number of scientists think that the origins of our nonverbal communication may be found in animal behavior. The presentation of an open palm as a gesture of friendship, for example, has been observed in chimpanzees by British anthropologist Jane Goodall. A young chimpanzee wanting to eat fruit in the presence of a more dominant member of the troop will turn his opened palm to the senior chimp before eating. The larger animal will then pat the palm of its subordinate as if in approval. Only after this display will the younger chimpanzee eat its food.

Frowning expressions can be seen when monkeys and other animals lower their eyebrows to examine something close to their faces. Some monkeys smack their lips and grin when they meet others of their species. The tense grin of nervousness may have originated as a protective tightening of the facial muscles, similar to the way an animal flattens its ears and narrows its eyes when threatened.

The idea that some of our most basic expressions are common to other species came from Charles Darwin. In *The Expression of the Emotions in Man and Animals,* published in 1872, Darwin ingeniously assembled a wealth of observations to demonstrate the long biological history of gesture and facial expression. He showed how a man's sneer of contempt "reveals his animal descent" in its similarity to the snarl of a dog. Both movements bare the canine tooth, although in man it is considerably blunter than a dog's. Darwin said, "When our minds are much affected, so are the movements of our bodies." For illustration, he described how the muscles of the face respond to sadness or anxiety. Muscles in the forehead and near the nose raise the inner corners of eyebrows and push them inward. These "grief muscles" form wrinkles on the forehead resembling three sides of a rectangle. Other facial muscles respond sympathetically, tightening and turning the corners of the mouth down.

Since Darwin's time, many investigators have studied the purposes and variations of human gestures and facial expressions. In many parts of the world, familiar gestures are interpreted differently. To greet their friends, Tibetans stick out their tongues; Bulgarians shake their heads in agreement. To form a ring with the thumb and index finger means "excellent" or "all right" in England and America, but in parts of Europe it can also signify "zero" or an insult. The gesture Americans use for "talkative" — bringing the thumb together with the other fingers — means "shut up" in France.

Throughout history, gestures have served both ceremony and art. Saluting superior officers is a common standard of military order. In the United States, political officeholders and witnesses in court raise their right hands to certify their probity. Paintings of Christ show his hand raised, the second and third fingers touching, the last two fingers turned downward. This sign of blessing is still used by the pope.

Traditionally, dances have been passed down through imitation and the memory of past performances. Now, however, choreographers can record their work in images that dance across a computer screen.

Charles Darwin said laughter, smiles and other expressions of emotion cross every cultural boundary. A tribeswoman from the highlands of New Guinea broke into laughter at her first sight of long, blond hair.

Many gestures now used without thinking originally represented precise ideas or actions. The handshake, our most common gesture of greeting, once indicated that the people clasping hands were unarmed. Crossed fingers signifying good luck probably first developed as a sign of the Christian cross to protect against evil. The low bow of introduction implies complete submission to authority by exposing the back of the neck, one of the body's most vulnerable parts.

In spite of the diversity of gestures in different cultures, the fundamental emotions are generally expressed identically. "Rage, anger, and indignation," Darwin noted, "are exhibited in nearly the same manner throughout the world." A number of scientists have disputed Darwin's views, but recent evidence vindicates his assertion. British anthropologist Desmond Morris has written, "The mouth kiss . . . contrary to popular opinion, is global in distribution." The kiss appears to have derived from mouth-to-mouth feeding between a mother and infant. Two Americans, Paul Ekman and Wallace Friesen, journeyed to New Guinea and photographed members of an isolated tribe that had no contact with Western civilization. They found that the tribe members used the same facial expressions for fear, happiness, surprise and other emotions that we use.

In conversation or even while sitting still, our faces are constantly moving, adjusting to new situations. Part of the reason why the classic films of Buster Keaton are so amusing is that his face never changed its mournful look regardless of the circumstances. For most of us, though, the face can mirror our inner thoughts, even those we might wish to conceal. Wrinkles show which muscles we habitually contract, whether from laughter or stress. The left side of the face is often said to reveal a person's true underlying character because its expression is molded by the right side of the brain, which some people claim gives freer rein to emotion. The right side of the face, subject to the more rational, less demonstrative left hemisphere of the brain, presents a more restrained countenance.

Paul Ekman, a professor of psychology in the medical school of the University of California at San Francisco, has another odd specialty. He can

Composite images of two left sides, below left, and two right sides of Richard Nixon's face reveal asymmetrical lines etched by muscles. The left side of the face is often said to exhibit emotions more intensely.

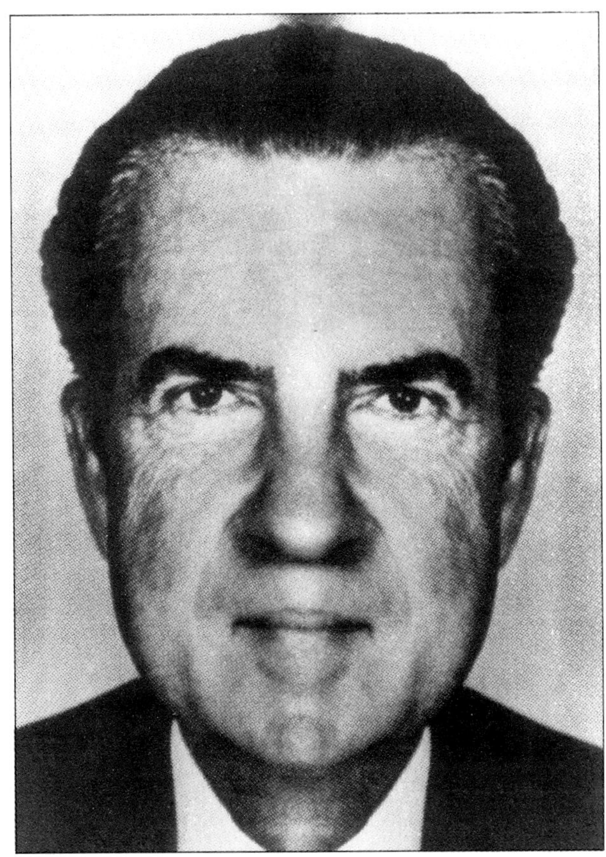

move the muscles of his face at will into about 7,000 different expressions. The average person's facial vocabulary consists of about 2,000. Ekman sees the face as a plastic instrument, giving shape to many different shades of feeling. A single event can trigger a variety of emotions that register on the face, sometimes for only a split second.

An understanding of gestures helped break through the soundless world of the deaf. Because many deaf people were also mute, little effort was made over the centuries to communicate with them. St. Augustine believed that the deaf could have no religious faith because faith was acquired only by hearing. But by the sixteenth century, deaf people were being offered education in several parts of Europe.

One of the greatest friends of the deaf was Charles Michel de l'Epée, a French priest born in 1712. He directed a school for deaf children and noticed that they communicated among themselves with hand signals. He mastered their gestures, developed new ones and became the first to spread a manual language among the deaf. In 1815 a young American, Thomas Gallaudet, went to Paris to study under de L'Epée's successor. When he returned to America, he opened a school for the deaf in Hartford, Connecticut and taught the manual method. Sign language has remained the predominant form of education for the deaf. Alexander Graham Bell, whose wife was deaf, taught "visible speech symbols," a form of lip reading, for a time. It was during experiments to amplify sound for the deaf that Bell invented the telephone.

Today, American Sign Language (ASL) is the leading form of manual communication among the deaf in the United States. It is not based on English. In fact, it cannot be understood by deaf people of Great Britain, who have a different system of manual communication. ASL has its

Overcoming blindness and deafness, Helen Keller used her hands to communicate with her teacher, Anne Sullivan. Movements are words in sign language, the gesture at bottom meaning "involve."

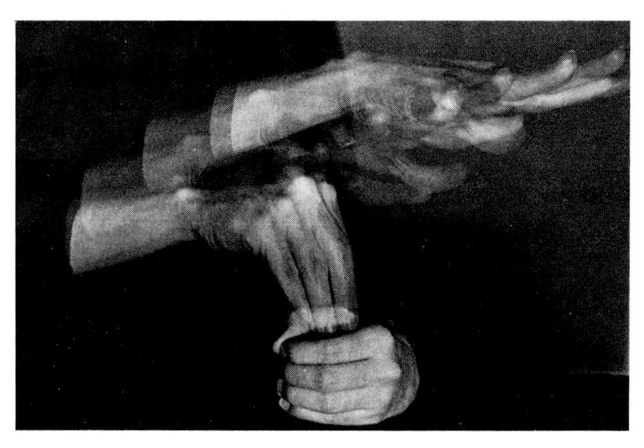

own grammatical structure and ways of word formation. The hands replace vocal cords, shaping the units of communication. Expressions complement the hand gestures of ASL.

Sign language is not a simple, rudimentary means of utilitarian discourse. It can be highly eloquent and more succinct than spoken English. Because its expression is visual, it has a richer vocabulary of sight than English. Many English phrases such as "to look forward to," "to do a double take" and "I didn't mean that" can be related with a single sign in ASL.

Since many words of ASL rely on a similar shape or position of the hand, humorous manual puns are possible. Sign language also lends itself to silent poetry and song: the hand movements become deliberate and rhythmical, giving the sense of metrical structure. Signs are sometimes elongated — like the vowels in songs — to fit the rhythm. Researchers who have filmed songs of deaf people in sign language have been able to count the number of beats per line and the pauses between stanzas. One group, the National Theater for the Deaf, presents plays, opera and poetry entirely in sign language.

Asian cultures have emphasized the relation between body and mind for centuries. Through careful muscular control, some practitioners of yoga and similar disciplines can accomplish feats that appear to defy medical explanation. Muscles are slowly stretched in yoga instead of flexed and released. Heartbeat and rate of metabolism slow down because the muscles are so relaxed. With these exercises and extraordinary self-control, yogis have lain buried alive for several days. In one case, a yogi voluntarily slowed his heartbeat so that it no longer registered on a monitor. After six days, he was uncovered with no apparent ill effects. Other yogis have manipulated their blood pressure and body temperature. Scientists are investigating yoga because of its power to control unconscious functions of the body.

For most of us, moods and movements are transitory; the body is in a constant state of flux, continually changing from one pose to another. Artists have long perfected their craft by drawing, painting and sculpting the human form. The best of them have endowed the body's strength

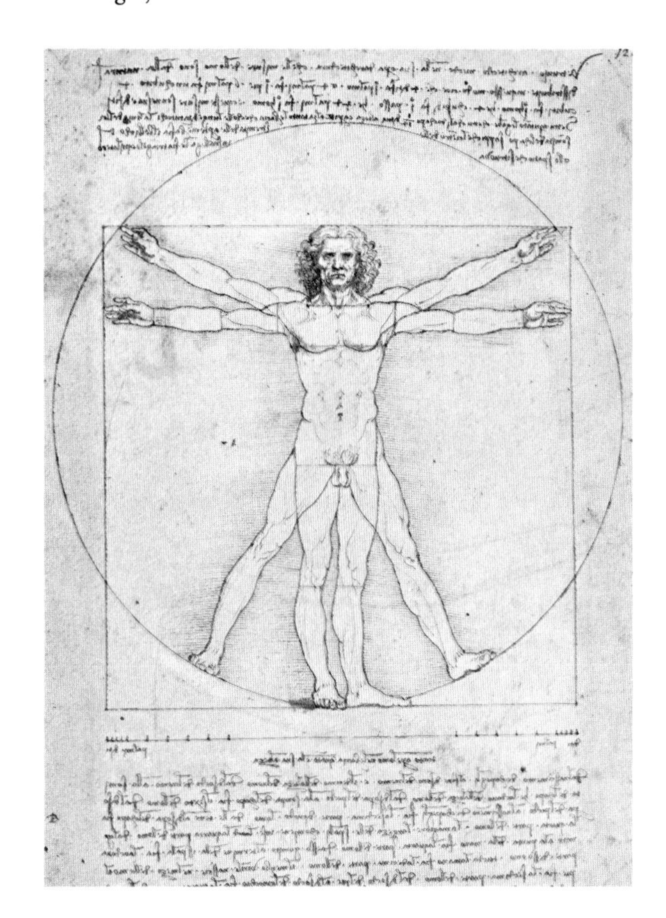

The symmetrical proportions of the body were proclaimed by Roman architect Vitruvius as a principle of aesthetic design. Leonardo da Vinci plotted human geometry in his drawing of Vitruvian man.

and grace with spiritual dignity. The words of Edmund Spenser suggest this elevation of the body out of the flesh: "For the soul is form, and doth the body make."

As with so many other principles of our civilization, understanding of the artistic potential of the human body began with the Greeks. They saw geometry in the shape of the body and portrayed their warriors and athletes in harmonious proportion. Polyclitus, a sculptor who lived in the fifth century B.C., developed a kind of muscular architecture of the male torso that soon became standard in Greek art. Known as the *cuirasse esthétique,* this formal depiction of a hero's body is marked by a characteristic horseshoe-shaped line where the torso meets the legs. This representation of the ideal body was so widely copied that it was even used to design armor breastplates.

In the first century before Christ, a Roman architect, Marcus Vitruvius Pollio, enunciated a doctrine of aesthetic principles drawn from human proportions. His idea of man as the perfect measure of all things influenced artistic thought for more than 1,500 years. Leonardo da Vinci and many other artists inscribed the Vitruvian man within a square and a larger circle to show the human body's harmonious geometry.

No one since the Greeks captured the human form so majestically as Michelangelo. The greatest draftsman of the Italian Renaissance, his work exaggerated the ideals of Greek art. His vision of the body was more sinewy than that of the Greeks. His drawings and sculptures reveal knotty muscles that undulate with an energy of their own within the contours of the body. He enhanced the idea of movement by placing his figures in twisting poses that emphasized muscular strain. In his ceiling murals for the Vatican's Sistine Chapel, he shattered artistic conventions by distorting human proportions for visual effect. His bulky, sculpturelike figures, in the words of British art historian Kenneth Clark, were powerful "mediators between the physical and spiritual worlds."

After Michelangelo, the most compelling interpretations of the human form were realized in nineteenth-century France by Edgar Degas and Auguste Rodin. Degas, a master draftsman, skillfully depicted women in a variety of ordinary tasks. He abandoned conventional ideals by celebrating the everyday beauty of women washing, combing their hair or exercising at a ballet barre. His paintings, in characteristic shades of pastel, reveal action suspended as if surreptitiously caught by a photographer. Eschewing the idealized forms of academic style, Degas saw in the body's simplicity a quiet dignity. His sculptures of young dancers gain an added touch of realism from tutus attached directly to the bronze.

Rodin's reputation was established through his original manner of introducing movement in his sculptures. Starting with a thorough knowledge of anatomy, he hired models to stroll around his studio rather than pose formally. He incorporated casual positions of the body into his sculptural figures but imbued them with a certain inner urgency. With their bulging muscles and bold gestures, they emanate strength, but often, as in the

To the Greeks, the body was an ideal form, a symbol of valor and youth. French artist Edgar Degas saw the body in a different light, though, without heroic grandeur or aggrandizement. His painting of a woman drying her neck, from about 1898, above, shows a simple, private action without embellishment. In contrast to Degas's calm view of the body, Michelangelo imbued it with restless strength. In this sketch from 1503, opposite, muscles bulge under the skin, imparting a powerful sense of movement even in an incomplete figure.

Burghers of Calais and *The Thinker*, a psychological heaviness hangs over them. They appear to be in epic struggle against destiny.

In 1912, at the age of 72, Rodin executed a small bronze of Nijinsky, the famous dancer, in his role of the mythical faun. The sculpture seems to resist gravity as much as Nijinsky's soaring leaps did. Rising on one leg, the other held to his chest, Nijinsky casts an enigmatic glance to one side, forever balanced in defiant grace. Although nearly fifty years apart in age, Rodin and Nijinsky had much in common: they departed from the confinements of convention to show the body in visually daring asymmetrical positions. By this time, the face of Western art had begun to change as society itself appeared to quicken its pace. The idea of movement was recognized as more than the pleasurable instinct at the root of dance and expression. It had become the metaphor of our century.

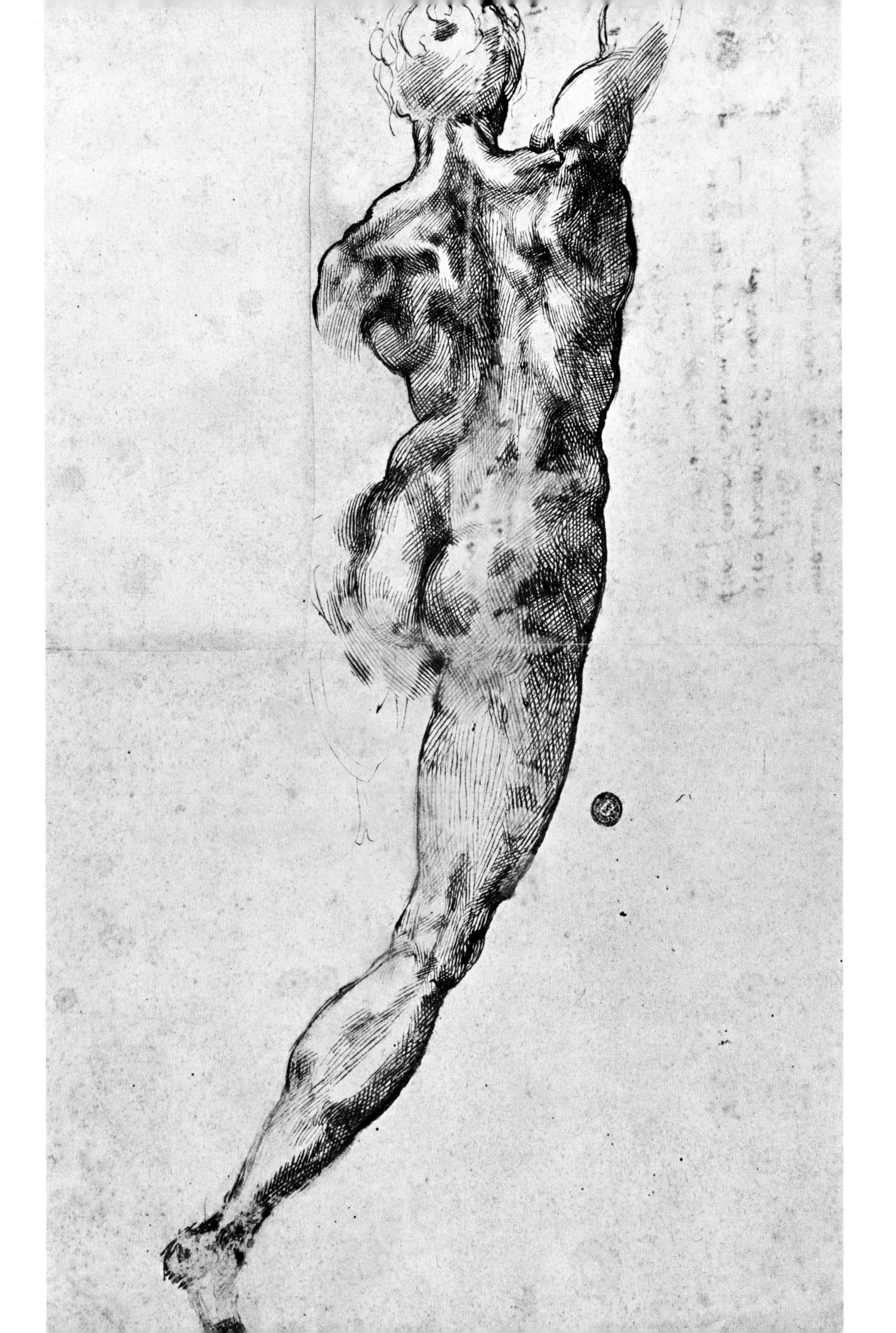

Chapter 6

Power and Prowess

Power and determination mark the body and gaze of Willis Reed, former center for the New York Knicks basketball team. The finest athletes in any sport possess a mixture of talent, discipline and competitive will that make their performances physical works of art.

I am inclined to believe," wrote neurologist Ernst Jokl, "that in all technological societies, sport represents the strongest remaining link between man and nature." Modern man has contrived to separate mind and body. For millions of people, the body merely supports the head, keeping it high enough to scan the top of a desk. From this elevation, the mind labors, with occasional help from the fingers and hands. In many occupations, it is the mind that wrestles, juggles and leaps. The body gets short shrift. Sports alter the balance of body and mind. Every sport mandates a few actions and prohibits innumerable others. Sports simplify. Within the circumscribed world of a sport, the mind is free to heed the body, its reactions and instincts.

For the individual, one of the thrills of sports is an indescribable feeling of rightness and effortlessness that accompanies a physical movement performed well. Sometimes, the golf ball seems to float off the tee; the golfer feels a swing not harder than usual but smoother, with no wasted motion, no errant twitch. Basketball players on a shooting streak feel they cannot miss, that the ball will automatically find the net. Baseball players on a hitting streak swear that the baseball looks bigger. But even great golfers drive into the trees. The best shooter goes cold. Every home run hitter has his slumps. The moments of effortless excellence cannot be commanded, which makes them all the more gratifying when they occur. Such a moment is one reason why people play sports. Sports writer John Jerome named a book after the phenomenon, *The Sweet Spot in Time.*

Sports are also social exercise. Between workouts, members of athletic clubs smoke, drink and converse, no matter how counterproductive the first two forms of relaxation may be. A round of golf can be a mixture of Sunday stroll, skill, petty gambling and high finance. The exact proportions depend upon the players.

First played by North American Indians, lacrosse, which they called baggataway, *was a wandering, often bloody battle among hundreds of warriors. Entire villages joined in, uniting sport, war and festival.*

The stakes in modern contact sports are higher, even threatening life and limb. They recall the same spirit that animated ancient spectacles and ritualized combats. Lacrosse, invented by North American Indians before the arrival of settlers from Europe, was sport, celebration and training for war. No one watching a professional football game can doubt that between the goal posts are the gladiators of the modern world. The clouds of cheers and hisses that roll down from the seats surely sound much the same in Los Angeles's coliseum as they did in Rome's. "Spectator sports," said former director of UNESCO René Maheu, "are the true theater of our day."

Many spectators openly enjoy sports for their violence. Boxing fans grumble when their fighter wins by a decision, hoping instead that he beats his opponent senseless. Even national honor rises and falls with athletics. Much of America celebrated during the 1980 Winter Olympics when the U. S. hockey team upset the Russians. The impetus behind the establishment in 1955 of the President's Council on Physical Fitness was a report by fitness experts Hans Kraus and Ruth "Bonnie" Pruden that America's youth was softer and feebler than Europe's. In studies conducted by Kraus and Pruden, almost 58 percent of American youngsters failed a six-part strength and flexibility test that more than nine out of ten European children passed. But for the woman jogging down the sidewalk, the teen-agers scrambling on an asphalt basketball court, and the rock climber resting for a moment in thin air, sports are something you do almost because you must. People have physical energy that demands expression. Through the jumps, kicks and swings of sports, people express themselves, as surely as they do through words or music. "The body, a true path to culture," wrote Albert Camus, "teaches us where our limits lie."

Helmets, pads and rules distinguish modern lacrosse from its progenitor. The University of Maryland, in red, and Johns Hopkins University are traditionally among the best lacrosse teams in the country.

Powering all this activity is muscle. Propelling an overweight body and a lawn mower across a yard is one kind of muscular work, although muscles work even at play. The muscles that pull a reader's eye across a page and the muscles that send a sprinter toward the finishing tape use the same kind of energy, an infinitesimal lightless spark released by the breakdown of adenosine triphosphate, ATP.

If all the ATP stored in a muscle at a given instant erupted simultaneously, the muscle would give one yank and sag in exhaustion. The movement of muscle is the result of an incessant instantaneous breakdown and reconstruction of ATP. ATP stores can activate a muscle for only a fraction of a second. As quickly as ATP fractures to free energy, it must be rebuilt, a process which requires energy. The release of energy from ATP is a small thing endlessly repeated. But the body eats and breathes to perform this chemical act.

Two methods employed by muscles to build ATP work anaerobically — that is, without oxygen. But the third method, the aerobic pathway, provides most of a muscle's ATP. The aerobic process takes place inside mitochondria, small sacs of enzymes and other substances scattered throughout cells. A mitochondrion's most vital job is to make ATP. Glucose and fatty acids, the building blocks of carbohydrates and fats, are its favorite foods. Aerobic metabolism is by far the most efficient way of making ATP in a muscle fiber, although not the speediest. The only byproducts of the process are water and carbon dioxide, which leaves the fiber along the same route oxygen enters. The aerobic pathway creates more energy than is needed for the formation of ATP. The excess, released as heat, is hardly wasted, however. It contributes to keeping the body at its most efficient operating temperature, roughly 98.6° F.

Cutting grass demands muscular effort and elevates the surface temperature of the body, as muscles burn fuel to liberate energy for contraction. High temperatures of excess heat show up as deeper reds in the thermograms, below. At the beginning of the job, left, the body is still idling as it generates a cooler yellow. Later on, the full throttle of exertion has generated the rich reds of higher temperatures.

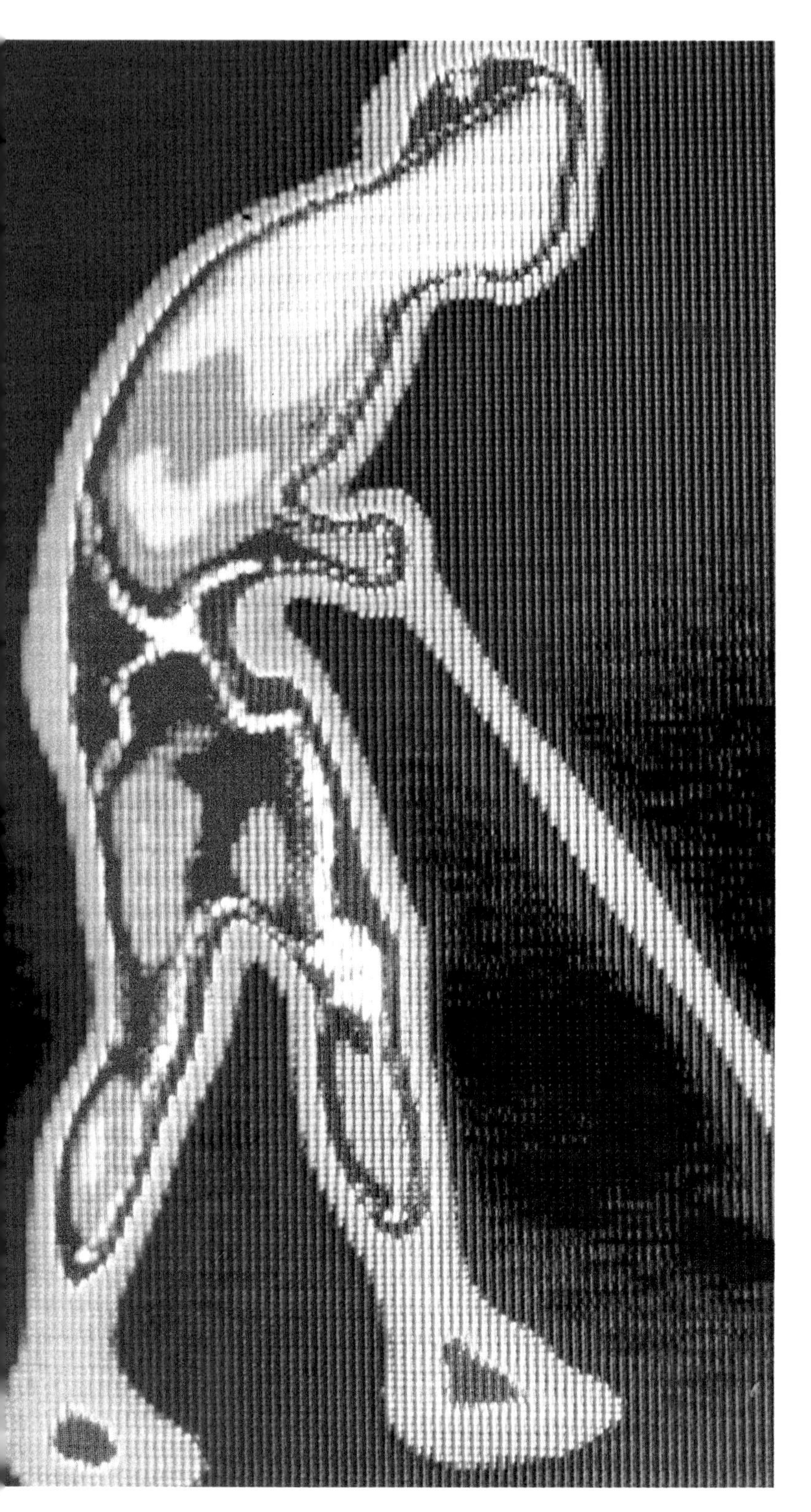

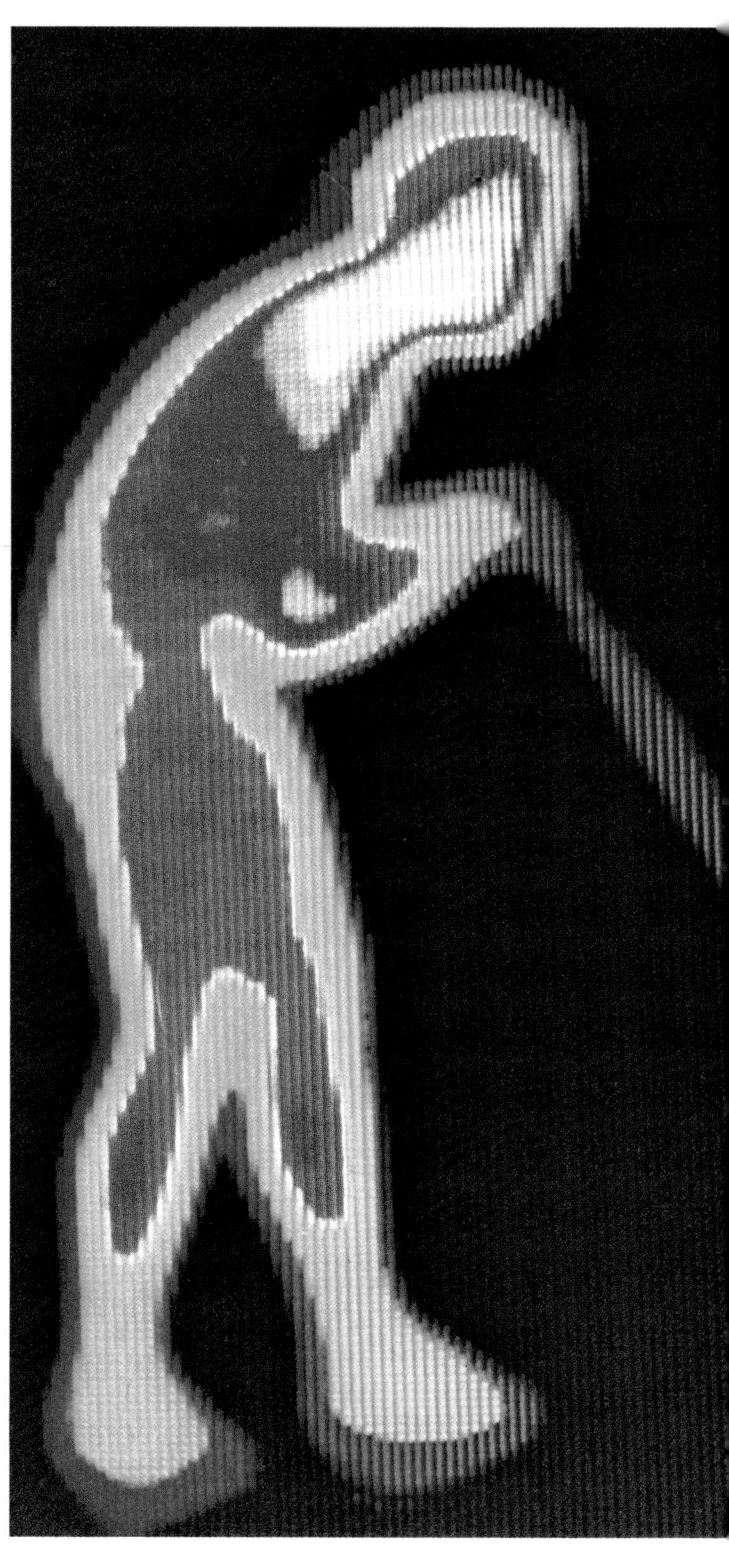

The aerobic pathway prefers fatty acids, if they are available. When a muscle rests, about two-thirds of its aerobic energy comes from fatty acids and one-third from glucose. This taste for fatty acids is called glucose sparing, a deference that muscles pay to the brain. Brain and nerve tissue can burn only glucose to create ATP. To save glucose for the nervous system, muscles burn fat whenever they can.

In resting muscle, only the aerobic pathway is in operation, supplying the minimal needs of limited motion. But more strenuous exercise prods the two anaerobic pathways into action. The first route depends on creatine phosphate, a substance similar in a way to ATP. Creatine phosphate and ATP are on opposite ends of a chemical seesaw. One releases energy and the other gains it. When a creatine phosphate molecule is carved by an enzyme into creatine and phosphate, it releases energy to make a molecule of ATP, whether the cell has any oxygen or not. This reaction occurs not in the mitochondria but in the midst of a muscle's protein filaments, near the site of contraction. Its greatest virtue is speed. In short bursts of intense activity demanded by sprinting, creatine phosphate is both the first source of energy summoned and the first depleted. The reserves of the compound in a muscle can only provide enough energy for about five to eight seconds of all-out exercise, a metabolic sprint. But the process allows muscles to react instantly without waiting for the heart, lungs and blood vessels to supply the extra oxygen that the aerobic pathway needs to work at full capacity. After it is used, creatine phosphate, like ATP, must be rebuilt. Paradoxically, the task depends on the energy liberated by ATP. Creatine phosphate lets the muscle rob Peter to pay Paul. Some of the ATP molecules created through the aerobic pathway power muscle fibers directly. But others help rebuild creatine phosphate. When the call for a burst of intense exertion comes again, the muscle is ready.

Anaerobic glycolysis is the second process supplying energy to muscle fiber without consuming oxygen. Glycolysis is the breakdown of sugar. The method allows muscles to borrow from their stores of glucose to make ATP while the more complex and efficient aerobic pathway gears up. Like the creatine phosphate pathway, anaerobic glycolysis works quickly. Enzymes, in a series of steps, transform glucose into pyruvic acid, and make ATP in the process. If a muscle fiber has enough oxygen, pyruvic acid slips into the mitochondria for further processing in the aerobic pathway. But when the aerobic pathway is clogged by a lack of oxygen, pyruvic acid cannot enter the mitochondria. In this case, other enzymes change pyruvic acid into lactic acid, which builds up in the cell and seeps into the blood.

The accumulation of lactic acid in muscle fiber makes the cell's internal environment more acidic. Higher acidity interferes with the work of the enzymes that break down glucose, so anaerobic glycolysis cannot proceed long without bogging down in its own by-products. Together, creatine phosphate and anaerobic glycolysis provide a muscle with only enough energy for about two or three minutes of intense exercise. A race over a medium distance, like 800 meters, can generate enough lactic acid to bring on exhaustion before the aerobic pathway has time to shoulder the burden of supplying energy.

Pathways to Power

In almost any exercise, the three pathways work in concert. Creatine phosphate and anaerobic glycolysis drive the muscles while the body makes the necessary adjustments to bring more oxygen to the muscles that need it. If the activity is not too strenuous at the outset, a mile or two of jogging as opposed to a 100-meter sprint, the anaerobic pathways soon shut down as the body achieves an aerobic steady state. At this point, some lactic acid has accumulated in the muscles and blood stream, but not enough to bring on exhaustion. Once the aerobic pathway is in full operation, it can use some of the energy it creates to transform lactic acid back into pyruvic acid, which mitochondria use directly to generate still more energy. The liver and muscles can also manufacture glycogen from lactic acid, another process that keeps the accumulation of lactic acid under control. Together, these processes check the build-up of lactic acid, forestalling exhaustion in moderately strenuous exercise.

Archibald Vivian Hill

Tinker of Living Machinery

A. V. Hill's first experiments ended badly. With a toy boat and nowhere to sail it but the bathtub, he tried to make a pond in the road by spitting. "The completeness with which that early experiment failed," he later wrote, "may be the reason why I recall it so vividly." A hen sitting on her eggs once drove him beyond the limits of his patience. The eggs, he decided, were being "unduly dilatory," so he opened a few of them at intervals to see how they were coming along. Hoping to succeed where the hen could not, he sat on the remaining eggs, ending the experiment. But his first pet might have been an omen of the illustrious career to come. Archibald Hill owned a white mouse.

Born in Bristol, England on September 26, 1886, A. V. Hill was a poor child, and brilliant. Hill's mother left his father when he was three. In his later years, Hill saw in himself the influence of his maternal grandfather — a bit of a dreamer, as Hill remembered him, a misplaced scholar or teacher, and a singularly unsuccessful woolens merchant. Excellence marked his academic career from primary school in Weston-Super-Mare to Trinity College at Cambridge University. A career in either

engineering or mathematics lay open to him. But his tutor at Trinity College was Walter Morley Fletcher, an eminent physiologist and one of the co-discoverers of lactic acid in muscle. Hill began experiments on the physiology of muscle contraction in 1909 and earned a Nobel prize for his work in 1922. His concept of an "oxygen debt," the ability of muscles to use oxygen to rebuild stores of energy after exercise, remains a cornerstone of exercise physiology. Muscles intrigued him for a lifetime. His *First and Last Experiments in Muscle Mechanics* was published in 1970, seven years before his death.

Hill's curiosity and the breadth of his intelligence often took him beyond the field of physiology and away from the laboratory, sometimes to his regret. Cambridge University elected him its representative to Parliament from 1940-45 and during the same years he traveled to India and the United States as a scientific adviser. At England's Air Ministry, Hill served on a committee which was, in his words, "the midwife of radar." His service to England's allies earned him France's Legion of Honor and the Medal of Freedom from the United States. In the midst of war hysteria, Hill's was an angry voice of compassion. He raged against the wholesale internment of refugees who fled a bloody Europe.

Possessed of the scientist's unflagging faith in the value of reason and perseverance, his bitterest foes were dogmatism and intolerance, scientific or religious. Hill saw logic in life, and wonder. "Truth can be pursued by many means," he wrote in *Living Machinery,* published in 1927, "and if we are right in our belief that a reasonable, consistent view of the world is possible, even though we ourselves may never conceive it, then there can be no ultimate conflict between those who pursue the truth, from whatever aspect, critically and carefully, without prejudice and with sufficient patience, reverence and imagination."

Figure skater Lynn Smith walks on a treadmill as scientists measure her ability to use oxygen. The amount of oxygen the body consumes during strenuous exercise is a valuable indicator of fitness.

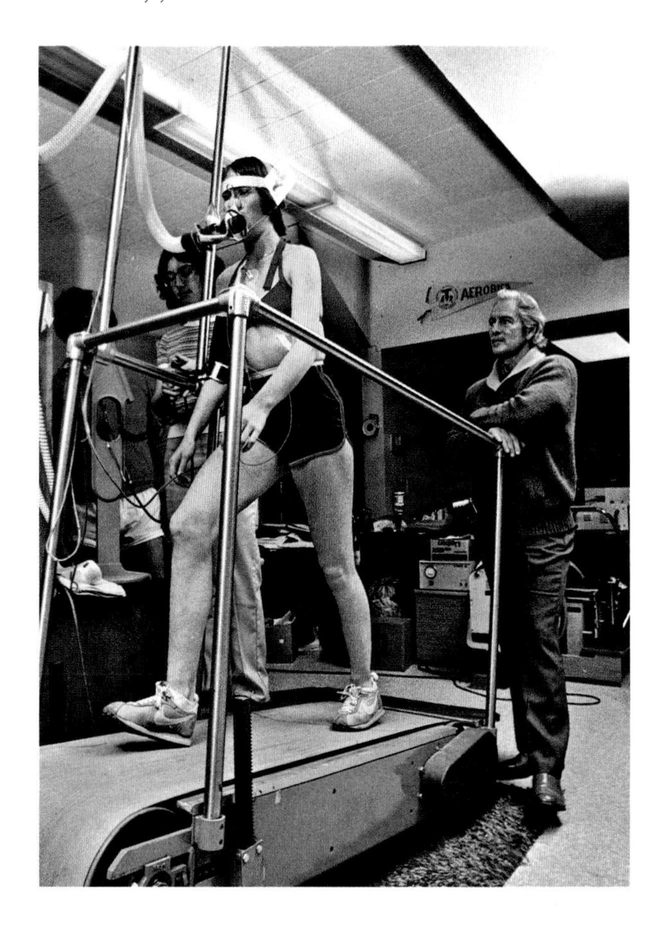

When using creatine phosphate or anaerobic glycolysis to begin a jog or fuel a sprint, the body borrows against its own resources. The labored breathing at the end of a race is the body's way of repaying this oxygen debt. The ability to run up an oxygen debt is one of the great assets of muscle tissue. Brain tissue would be long dead if forced to go without oxygen for as long as some muscles do. Muscle cells contain myoglobin, an oxygen carrier similar to hemoglobin in blood. Myoglobin is an oxygen bank, releasing oxygen slowly as the muscle needs it, even when little oxygen is reaching the cell from the blood.

Repaying the oxygen debt serves two main purposes, the replenishment of creatine phosphate supplies and the disposal of lactic acid. Even after exhausting exercise, just two or three minutes of heavy breathing allows the body to produce enough energy aerobically to rebuild stores of creatine phosphate. To cleanse tissues of lactic acid, the body can send it through the aerobic pathway to generate energy or through the liver and other tissues to reconstruct glycogen. A sprinter's oxygen debt is repaid an hour after a race, even though he has burned glycogen stores that must eventually be replaced. The process leaves him with energy in the bank, so to speak, and ready to sprint again.

In the marathon, a race of just over twenty-six miles, a runner crossing the finish line does not feel the sluggishness of lactic acid accumulation. Instead, the marathoner experiences a different and more complete exhaustion. When the body exercises for many minutes at the peak of aerobic capacity, exactly the way a skilled marathoner would run, it steadily burns up the muscles' stores of glycogen, draws heavily on glycogen stores in the liver and taxes the available supplies of fatty acids as well. The process is almost totally aerobic. The level of lactic acid in a distance runner's blood might be only two or three times normal, compared to that of a wrestler, which might reach twenty times the normal level. The exhaustion a marathoner feels as he crosses the finish line comes from the near total depletion of the body's stores of fuel. Recovery from this kind of exertion demands more than deep breathing. Replenishing stores of fats and glucose calls for food and rest. A marathoner who runs too many races in succession, or even a sprinter who races 100 meters too often in a day, too many days in a row, runs the risk of constant fatigue. Endurance athletes need several days between races and a diet high in carbohydrates to replace liver and muscle glycogen.

Not everyone wants to be a marathoner. Nor does everyone relish exercise. Senator Chauncey Depew of New York once quipped, "I get my exercise acting as a pallbearer to my friends who exercise." Robert Hutchins, former president of the University of Chicago, is supposed to have once remarked, "The secret of my abundant health is that whenever the impulse to exercise comes over me, I lie down until it passes." But the relentlessness with which superb athletes push their bodies illustrates a crucial point about muscles. They exist to be used. Looking at an arm just freed from a cast after two months of

Eric Heiden, the world's best speed skater, needs anaerobic and aerobic energy — speed and endurance. Heiden holds Olympic gold medals in the 500-meter, the 10,000-meter and every race in between.

Cross-country skiing is excellent aerobic exercise. Arms, legs, lungs and heart must all labor mightily to propel a skier across the snow. Ski touring, a combination of cross-country skiing and winter camping, is even more demanding. This skier faces the fierce winter of Crater Lake National Park in Oregon.

imprisonment is all the evidence needed. Left alone, muscles atrophy. Challenged, they grow.

Exercise reveals one of the great virtues of the human body. "If you will only ask the organism for more," says John Jerome, "it will eventually respond." Even moderately dedicated training encourages changes in the body that go far beyond bulging biceps. Exercise helps the whole body, inside and out.

Building Better Muscles

A general benefit of exercise is good muscle tone. Within many muscles, especially those that help maintain posture, some fibers must contract most of the time. When standing, the body sways gently around a perfectly upright position, responding to subtle contractions of muscles of the legs and trunk. Muscles exercised regularly do a better job at everything, including holding the body erect. Good muscle tone also implies a readiness to move, a kind of muscular alertness and resistance to fatigue.

The human heart, itself a bag of muscle, is perhaps the greatest beneficiary of exercise. The demands strenuous exercise places on the heart can be enormous. From a resting volume of about 5 quarts a minute, a young man's heart can pump 25 to 30 quarts a minute. An even more remarkable increase occurs in the lungs. From about 7 quarts of air a minute at rest, a body exercising to its limits demands enough oxygen to pump 150 quarts of air per minute into the lungs. Strenuous exercise does not overtax the heart and lungs, but strengthens them. As regular exercise challenges the heart, it grows stronger, draws more blood and oxygen through the coronary arteries and increases the amount of blood it pumps with every pounding stroke.

A larger stroke volume means that the heart can pump the same amount of blood with fewer beats. At rest, the hearts of some world class athletes beat forty times a minute, compared to the average person's sixty to a hundred beats. Roger Bannister, the first man to run a mile in under four minutes, had a resting heart rate of thirty beats per minute. Over time, the coronary blood vessels send out new capillaries to feed a laboring heart. Additional capillaries help prevent heart

DESCENTE

attacks by increasing the number of routes by which blood can bypass a blocked artery. The inner diameter of a coronary blood vessel widens with exercise, which may discourage deposits of fat from obstructing the flow of blood.

Inside the muscle fibers themselves, other changes take place. Exercise a flabby, ignored muscle, and it gets bigger and in a sense, better. Muscles enlarge by a process called overloading. Mild exercise will maintain them. But subjected to unaccustomed labor — and for truly sedentary people almost any exercise can be considered out of the ordinary — they will grow.

The larger and more numerous capillaries bringing blood to exercising muscles are part of this growth. A body builder's huge muscles bulge partly with blood vessels. Sufficiently overloaded, myofibrils — the protein cords inside muscle fibers — increase in size and number. A muscle's strength depends on this dual increase. The greater a muscle's bulk, the stronger it is. Contrary to popular belief, bigger muscles are not less flexible. Often the opposite is true, since building strength can also increase flexibility. People with big muscles are not muscle-bound. The term better applies to people so out of shape that they are bound by their flabbiness.

Bigger muscles are often better in the sense of being more efficient. Regular, strenuous exercise improves a muscle's ability to produce energy, anaerobically and aerobically. After a few weeks of exercise, fibers will show more myoglobin and bigger, more abundant mitochondria. Inside the mitochondria, there is an increase in enzymes that help transform fatty acids and glucose into aerobic energy. Stores of ATP and creatine phosphate are greater in fit muscles. Strong muscles stockpile more glycogen and fat than do weak ones, and they burn fuel more efficiently as well. Fit muscles resist fatigue.

Psychiatrists and psychologists report that regular exercise can limit the need for medication and improve progress in therapy. Some of the jogging faithful swear that the sport has changed their lives. The psychological benefits of exercise, although more difficult to analyze, are just as real. Anxiety, boredom, frustration, tension and anger, like glucose, are all to some extent burned off with a little exercise. There is even some evidence for a biochemical thrill from running called "runner's high." Repetitive, strenuous exercise may persuade the brain to release small amounts of chemicals called endorphins, which ease or even erase pain. Brain chemicals aside, exercise breeds hope. A fat, lazy body coaxed into shape is living proof, at least to the person who owns it, that human beings can change for the better.

No exercise is inherently better than another. It all depends on what people seek from it. Some people want strength, others endurance, others both. Many people simply want to have fun and feel fit. But the division between strength and endurance mirrors a larger split that runs through all kinds of exercises and sports and down into muscle fibers themselves.

Cords of Strength and Stamina

Skeletal muscles are composed of two kinds of fibers, slow-twitch and fast-twitch. Fast-twitch are the fibers of anaerobic energy, the fibers of strength and power. They contract quickly, yielding short bursts of energy. A motor nerve and the muscle fibers it controls are together called a motor unit. Fast-twitch motor units have many fibers, which compounds the force of their individual contractions. Slow-twitch fibers have a steadier tug, and, generally, fewer of them make up a single motor unit. They produce most of their energy aerobically and tire only when their supplies of fuel are gone. Slow-twitch fibers are the cords of endurance.

In the human body, where many distinctions blur, these two kinds of fibers do not behave quite so distinctly. All muscle fibers have the ability to make energy aerobically and anaerobically, although not to the same degree. All are strong and steady, but some are designed primarily for strength and others for steadiness.

Slow-twitch fibers house many mitochondria that hold the enzymes driving aerobic metabolism. They are surrounded by dense capillary beds and draw more blood and oxygen than fast-twitch fibers. Their rich blood supply, plentiful myoglobin and thick intracellular fluid make slow-twitch fibers appear darker under a microscope. Slow-twitch fibers are dark meat.

Dark slow-twitch fibers and lighter fast-twitch fibers mingle in the micrograph below. A different staining technique, bottom, reveals a subtler distinction. Some fast-twitch fibers, such as the large fiber at the center of both micrographs, tire more easily than others. To the right of that fiber are two slightly darker fast-twitch fibers with greater aerobic capacity, called "fast-twitch resistant to fatigue."

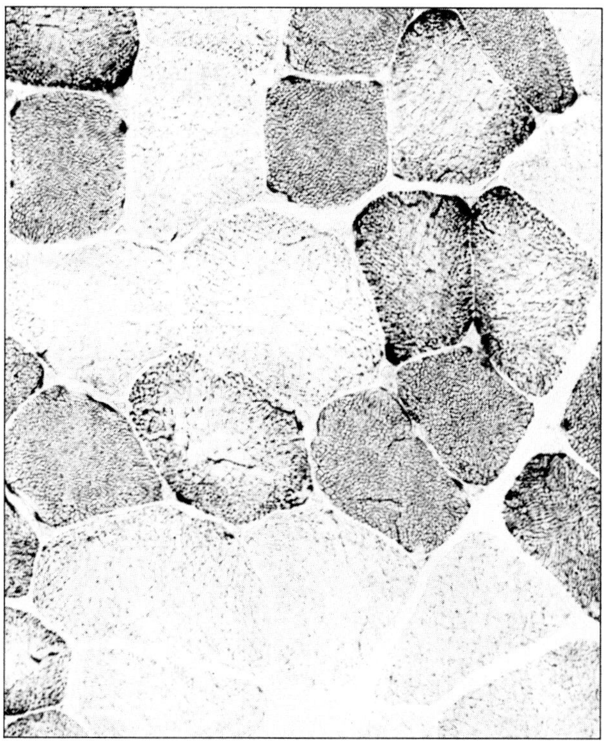

Lacking the dense webbing of capillaries, fast-twitch fibers are lighter in color — the white meat of muscle. Slightly larger than slow-twitch fibers, they have a greater capacity for anaerobic glycolysis. Because they provide explosions of energy, fast-twitch fibers are recruited most heavily for brief, intense exercises — sprinting, weightlifting, swinging a golf club or putting a shot. Their dependence on anaerobic energy leaves them vulnerable to the by-products of their own strength. An accumulation of lactic acid drives them to exhaustion quickly.

All skeletal muscle possesses fast-twitch and slow-twitch fibers in varying proportions. The exact mixture varies from person to person and from muscle to muscle in any individual. Many scientists are coming to suspect that the distinction between fast- and slow-twitch fibers may not be as absolute as once believed. Muscles may also contain a variety of fast-twitch fiber that is richer in mitochondria, blood supply and aerobic capacity than most fast-twitch fibers, a variety called "fast-twitch resistant to fatigue."

As a rule, however, the division between fast- and slow-twitch fibers remains valid. Athletes around the world depend on it, because different exercises tax not only different muscles, but different fibers within a muscle. Jogging tests the legs and strengthens them. But it does not give them the strength a weightlifter or a wrestler needs. Specific sports demand specific exercises.

Endurance exercises promote overall fitness. The President's Council on Physical Fitness and Sports enlisted seven medical experts to rank the contributions of several sports to physical well-being. The panel declared jogging, bicycling and swimming, in that order, most beneficial. Golf, softball and bowling were last.

A race called the Hawaii Iron Man Triathlon is held every year for the pleasure of men and women chasing the horizons of their own physical ability. The triathlon consists of a 2.4-mile swim, a 112-mile bicycle race and a 26.2-mile marathon — all in one day. Actually, the title "Iron Man" is a bit misleading. In 1979, a woman, Harvard law student Lyn Lemaire, placed fifth.

The price athletes pay for their efforts, from the weekend golfer to the Triathlon champion, is

Even in ancient Greece, athletes had trainers and attendants. An athlete doffs his cloak, right. Another oils his body before training. At left, an attendant pulls a thorn from the foot of an injured competitor.

soreness, pain and injury. Even joggers sacrifice themselves to sport. Richard Mangi, Peter Jokl and O. William Dayton, authors of *The Runner's Complete Medical Guide,* declared, "Runners are the fittest group of sick and injured people in the world." Muscle soreness, specifically the variety that strikes within two days of a workout, is probably the result of minor muscle injuries that encourage swelling and pain. The best treatment for this kind of stiffness is preventive. Coaches, athletes and scientists universally proclaim the virtues of warming up before exercise and cooling down afterward to prevent pain and stiffness. Calisthenics or a brief jog raises the temperature inside muscles, draws more blood toward muscle fibers and helps enzymes work more efficiently. Cooling down with a similar regimen after a hard workout helps clear lactic acid and excess fluids from muscles and pump them into the blood stream, easing both stiffness and fatigue.

Muscle strains, also called pulled muscles, are slight tears in muscles or tendons. The damaged area is often too small to bleed. But even tiny tears in muscles can send the fibers into painful spasms, which, untreated, can persist for days or weeks. Muscle strains are usually the result of too much tension on the muscle. A blow to a muscle seldom causes a strain. When strained muscles heal, scar tissue replaces the injured muscle fibers, weakening the muscle somewhat, but rarely enough to impair an athlete's performance. The hamstring, which stretches down the back of the leg from hip to calf, can be an exception to this rule. Hamstring strains often recur and can eventually force some athletes out of top level competition.

A sprain damages one or more ligaments, which connect bones and help keep joints strong and stable. A misstep, collision or accident is usually the cause of a sprain. The most painful sprains are not always the most severe. A completely torn ankle ligament might cause less pain and swelling than one slightly torn or stretched. Ligaments are more difficult to rehabilitate than muscles. Complete ruptures require surgery.

A number of athletic injuries occur from overuse. Tennis elbow, usually the result of a bad

Levers, a sliding seat and assorted machinery make up the Sargent inometer, a training device for rowers and bicyclists that appeared in Robert Tait MacKenzie's Exercise and Education in Medicine.

backhand that overstresses muscles and tendons of the forearm, is the most obvious example. Shin splints, bruised heels and stress fractures are a runner's most common injuries from overuse. The muscular or skeletal problem that causes the pain of shin splints still eludes doctors. Rest is the only cure. Stress fractures generally occur when converts to jogging attempt too many miles too soon. Bones will strengthen and grow in response to strenuous exercise, but the process takes a little time. Demanding too much too soon of the leg or foot sometimes produces a painful microscopic crack across the bone.

Careers Cut Short

Basketball players and high jumpers can suffer from jumper's knee, an inflammation of the tendons around the knee joint. Little League elbow afflicts a few young baseball players who overwork their throwing arm. Some swimmers and professional baseball players have their careers cut short by rotator cuff injuries. The rotator cuff is a band of four muscles that helps hold the shoulder joint together and powers the arm. The extraordinary stress that throwing thousands of pitches or swimming hundreds of miles puts on the joint can slowly erode the cuff's muscles. For professional athletes, the injury is serious indeed. No major league pitcher has ever recovered from rotator cuff surgery and returned to his earlier form. The first treatment for injuries of overuse is rest. Once the injured muscle or tendon has had a chance to heal, a better pair of jogging shoes, a few tennis lessons or a little moderation usually prevents recurrences of the injuries.

Physicians have studied the benefits and risks of exercise since at least 800 B.C., the date of an Indian medical text, the *Ayur-Veda,* which prescribes exercise for rheumatism. Since the age of Aristotle, in the fourth century B.C., men have probed the mystery of muscle and motion, seeking to discover exactly how people move.

Ancient Olympians trained under *gymnastai,* coaches and trainers. The most famous of the Greek *gymnastai,* Herodicus, was the teacher of Hippocrates and an ardent supporter of exercise as physical rehabilitation. Hippocrates himself, anticipating modern psychologists, prescribed exercise for mental disorders. Like modern-day trainers, the *gymnastai* of old complained about the laziness and softness of their athletes. And like modern athletes, some apparently broke training to indulge their huge appetites. Two gluttonous tales survive of Milo, a remarkable athlete from the city of Croton, a Greek colony in southern Italy. Milo was Olympic champion in wrestling six times. According to one story, he once killed a heifer with one blow of his fist and ate the meat of the entire animal. Another story credits him with downing nearly twenty pounds of meat, twenty pounds of bread and eighteen pints of wine in a single day.

By the Renaissance, daily exercise was a part of the curriculum at some schools. The seventeenth and eighteenth centuries gave birth to brilliant studies on orthopedic medicine and the movement of animals and humans. Nevertheless, it was only in the twentieth century that modern science began to grasp the inner workings of man in motion. The study of exercise physiology blossomed in the work of Canadian sculptor and physician Robert Tait MacKenzie, whose book *Exercise and Education in Medicine* appeared in 1909, and reached full flower in the work of British physiologist A. V. Hill in the 1920s.

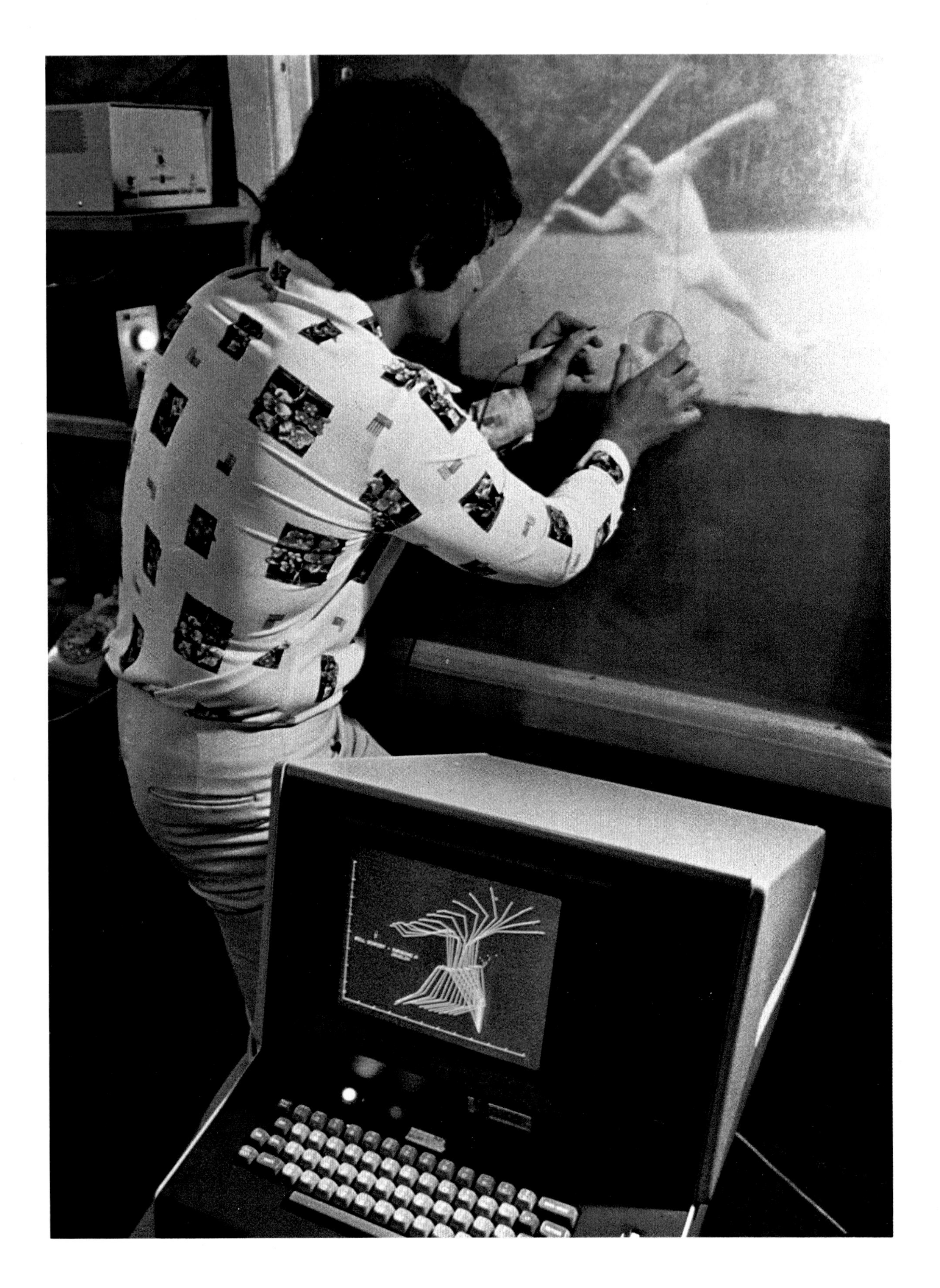

Contemporary sports medicine embraces the physiology of exercise, physical fitness, rehabilitation, orthopedic surgery and several other fields. Among them is biomechanics, the study of the mechanical forces at work in the movement of living bodies. The discipline traces its roots to Aristotle and other ancient thinkers. But it has matured today into a complex science that relies on slow-motion photography, computers and other sophisticated devices. Biomechanists study movements, forces and postures that may appear for only fractions of a second in the course of a single motion. But by analyzing a stride, throw or swing in its mechanical details, they learn about the intricacies of human motion, the special skills and needs of athletes and the flaws of many kinds of equipment and advice.

At the biomechanics lab of Pennsylvania State University, scientists study everything from bicycling and running to golf and jogging shoes. By recording the force with which a runner's foot strikes a sensitive metal plate, biomechanist Peter Cavanagh learned enough about the mechanics of the human stride to prompt the redesign of millions of jogging shoes sold in the United States every year. Biomechanics, in this instance, led to changes like better cushions across the sole of the shoe and, quite possibly, to less pain and fewer injuries for many runners.

An ultimate goal of biomechanical research is to remedy subtle flaws in the performance of athletes. By analyzing slow-motion films of a runner's stride, biomechanists can use a computer screen to generate a stick figure completing a stride in stark simplicity. Ideally, careful scrutiny of the runner's motion on the computer might reveal a stride slightly long, a foot turned out or some other nearly invisible error.

Biomechanics offers no short cut to excellence. The motions of athletes are highly idiosyncratic, so good advice to one runner might be useless to another. Nor can biomechanics compensate for poor training, a lack of competitiveness or a weak will. Nevertheless, biomechanist Gideon Ariel of California's Coto Research Center has convinced at least two world class athletes of the value of taking a scientific approach to their sports. After working with Ariel, discus thrower Mac Wilkins

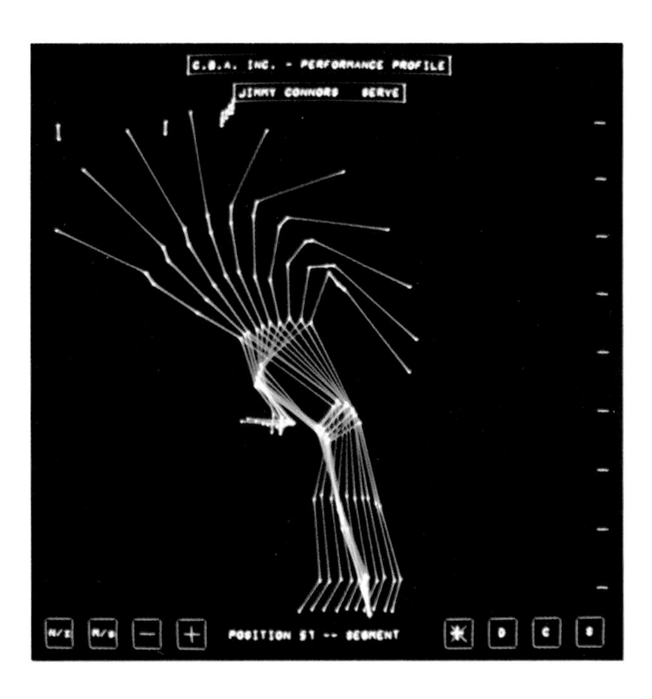

Biomechanist Gideon Ariel traces the leg of javelin thrower Bill Schmidt as part of his analysis of Schmidt's throw. By observing slow-motion films of an athletic performance and reproducing the athlete's motion with a computerized stick figure, biomechanists hope to help athletes improve their technique, and thus their performances. A biomechanical stick figure, above, recreates a movement many tennis players wish they could imitate — Jimmy Connors's serve.

Biomechanical studies at Harvard showed that a runner's head and hips stayed nearly level on both a hard and soft track, top and bottom. But a track's springiness does influence stride length and speed.

improved his longest toss from 219 feet, 1 inch to 232 feet, 6 inches, and later earned a gold medal in the 1976 Montreal Olympics. Ariel's techniques also helped shot putter Terry Albritton heave the sixteen-pound ball 71 feet, 8 ½ inches in 1976 to set a world record, 2 feet farther than Albritton had thrown before.

The methods of biomechanics take scientists beyond equipment and performances to the ground beneath an athlete's feet. At Harvard University, engineers Thomas A. McMahon and Peter R. Greene used biomechanical principles to design a "tuned" track for the Harvard track team. The track is tuned to the human step. As a runner plants his foot on the track, some of the energy he imparts is stored in the track and returned to his foot for his next stride.

McMahon and Greene predicted that their tuned track, a surface of polyurethane with a wood rather than cement or concrete substructure, should improve runners' times by 3 percent. In the first year the track was open, the 1977-78 track season, Harvard's runners improved their times an average of 2.91 percent. Runners from other schools gained about a 2 percent advantage. But the tuned track's greatest virtue may be that it allows runners to train harder without straining their legs and risking injury. The tuned track is faster and gentler.

In their search for a competitive edge, athletes scour other areas of sports science, especially nutrition and pharmacology. Many athletes resort to special diets, which sometimes include astounding doses of proteins, vitamins, minerals and other more exotic substances. Some weightlifters routinely take one or two hundred vitamin and mineral pills a day during intense training.

Glycogen loading is another dietary manipulation practiced by some athletes. The process requires a week to complete. First an athlete must work to exhaustion to deplete the stores of glycogen in his muscles. For the next three days, he continues to exercise strenuously and eats mostly fats and protein with very few carbohydrates. Over the following three days he rests, eating a high carbohydrate diet. At the end of the cycle, an athlete can raise the glycogen in his muscles as much as 300 percent. Whether this somewhat radical procedure improves performance is still unknown. An athlete's ability to store glycogen may be less important than his capacity for burning it efficiently. Superb athletes, who already store more glycogen than the average person, may gain little from glycogen loading. And weekend athletes who load up on glycogen to enter a marathon or prepare for a big tennis match may gain no endurance. Moreover, muscles packed with glycogen store water. An athlete can gain two or three pounds of water along with the extra glycogen, which may produce a feeling of stiffness.

Caffeine and Competition

One recent study at the Human Performance Laboratory at Ball State University in Muncie, Indiana found that the caffeine in approximately two-and-a-half cups of coffee, ingested about one hour before a race or match, helps the body burn fatty acids instead of glycogen. In theory, this delays the time when the depletion of glycogen supplies brings on exhaustion. In marathons and other endurance sports, conserving glycogen could conceivably cut a runner's time by several minutes. But like glycogen loading, caffeine is no guarantee of success. The best runners burn more fatty acids than the average person anyway, so they might have the least to gain from gulping down two or three cups of coffee before a race. In addition, competition, like caffeine, makes many people nervous. Runners who tend to climb the walls before big races might be better off without additional stimulation.

Once the quest for the competitive edge goes beyond caffeine and carbohydrates, the techniques become progressively more unsavory. The outstanding Finnish long-distance runner Lasse Viren was accused, without proof, of blood doping for the 1976 Olympics. Blood doping is the process of removing a pint or so of blood several weeks or months before a competition, storing it and reinfusing it back into the vascular system a few hours before an event. The more blood, the more red blood cells, and, theoretically, the more oxygen delivered to the tissues. The accusations against Viren may have stemmed solely from envy. He won the 5,000-meter and 10,000-meter

Aspiring strong men at New York City's Gladiator Gym lift weights to build muscle mass and strength. Weightlifting, which usually brings forth more grimaces than smiles, is a form of isotonic exercise.

runs in the 1972 and 1976 Olympics, a performance that left his competitors searching for any reason for his success. Scientific experiments on blood doping have produced conflicting results. Although discouraged, blood doping has not been prohibited in Olympic competition. No reputable sports physician encourages it. But with simple precautions, the process is not dangerous, and some athletes use it to find that extra spark of energy that might win a race.

To push themselves beyond their normal limits, some athletes resort to drugs. Alcohol is an innocuous example. Many target shooters take a drink to steady their hands before matches. At the other extreme are athletes who use amphetamines or cocaine to give themselves more endurance or a psychological edge. The use of these drugs is universally decried. Abuse can lead to addiction, and in the case of amphetamines, circulatory collapse and death.

Somewhere between amphetamines and alcohol on a scale of disrepute and risk lie steroids. Steroids are a class of chemicals that include bile acids, some naturally occurring drugs such as digitalis and many hormones. Among them is the male hormone testosterone, which influences the human body in two ways, androgenically and anabolically. Testosterone's androgenic effects induce secondary sex characteristics, including body hair and a deep voice. Its anabolic powers promote growth, particularly muscular growth. Anabolic steroids are synthetic drugs that mimic the growth-promoting effects of testosterone, with only a mild androgenic influence. An athlete on a strength-training program who has plenty of protein in his diet may gain an edge by taking steroids, although this has never been proved. Their use among top weightlifters, shot putters, discus throwers, football players and other athletes who need strength is, to say the

Moments in an arm wrestling match when opponents battle and growl each other to a standstill are examples of isometric exercise. Their muscles are strained to the limit, but motionless.

least, widespread. The International Amateur Athletics Federation (IAAF) and the International Olympic Committee, which together preside over all important international track and field competitions, have declared steroids illegal. At the Olympics, athletes must submit to elaborate tests aimed at detecting the presence of steroids or be excluded from the games. The IAAF checks athletes randomly at its major events. Steroids taken orally leave no traces in the body after an athlete stops taking them for two weeks to two months, depending on the individual. Injected steroids take slightly longer to disappear. Within a day or two after steroids are discontinued, their effects start to wear off; athletes begin to lose strength and weight. But many athletes believe that the power of the drugs to add pounds and muscle is great enough to offset the loss.

Studies of steroids have uncovered potentially serious long-term side effects, including liver damage, hepatitis, and lower sperm production. But athletes still use them. The IAAF stripped American discus thrower Ben Plucknett of his world record in 1981 and banned him from all international competition when tests revealed traces of steroids in his urine. In 1977, after traces of steroids cost him the right to compete internationally, Finnish discus thrower Markku Tuokko voiced what is probably a widespread opinion among athletes whose sports demand strength: "Without hormones, there is no way one can reach international top-level competition and make new records these days." Few believe disciplinary measures will stop the use of steroids. The competitive pressure is too great.

Whatever advantages biomechanics, diets or drugs may bring athletes, the basis of improving athletic performance will always be training. Strength is good for sports, and muscles grow stronger by being overloaded, stressed to their limit. Top athletes in many sports spend a good portion of their time training to increase strength. Three distinctly different kinds of muscle contractions — isometric, isotonic and isokinetic — build muscle strength.

Isometric, from the Greek for "constant length," is a contraction in which the muscle fights an immovable force so that it tightens without changing its length. If a person stands in a doorway and pushes against the sides of the frame, his muscles are contracting isometrically. Isometrics can also pit one muscle against another. Placing one's hands palm to palm and pushing is an isometric exercise. Isometrics requires no equipment. Anyone can perform the exercises anywhere. Isometrics enjoyed great popularity in the United States in the 1950s after two German scientists, Theodor Hettinger and E. A. Müller, reported that holding an isometric contraction for six seconds, once a day, five days a week, could increase strength about 5 percent per week. Since the muscle in question does not move, however, isometrics tends to build a muscle's strength only at the angle at which the exercise is done. When a person puts his hands under a heavy table and lifts with his elbow bent at a 90-degree angle, he builds strength in his biceps, but mostly at that particular angle. Strengthening all the muscles of the upper arm demands repeated isometric contractions at many different angles.

Weight training and many simple exercises such as chin-ups and sit-ups take advantage of isotonic contractions. Although isotonic means "constant tension," the term is a misnomer. As a muscle straining against a weight moves through a range of motion, tension changes as the angle of motion changes. In a chin-up, the angle of the elbow joint changes nearly 180 degrees and the tension on the biceps changes constantly through the range of motion. The beginning and end of a chin-up are the most difficult stages, because at those angles the muscles of the arms cannot generate as much tension.

Many experts believe that weights and barbells are not enough for building strength. To add powerful pounds of muscle most efficiently, hundreds of colleges and professional sports teams use special weight machines. Among these, the most familiar is the Nautilus machine. Begun in 1970, Nautilus Sports/Medical Industries is an athletic empire, supplying weight machines to twenty-seven of America's twenty-eight professional football teams and more than a thousand colleges and athletic clubs. Nautilus machines, like other weight machines, consist of chains and pulleys connected to a stack of weights. In a Nautilus machine, at least one wheel is irregularly shaped, like the spiral shell of the nautilus, a sea mollusk. The special wheel adjusts the resistance muscles overcome at various points in the exercise, rectifying a problem associated with isotonic exercises. Nautilus machines are unquestionably effective. Unlike barbells, however, the average person cannot buy a set and install them in his basement. Because each machine works only a particular group of muscles, a complete Nautilus center requires about a dozen machines, altogether costing thousands of dollars.

Machines designed to build strength through isokinetic, or "constant speed," contractions can also be expensive. But like Nautilus machines, they offer advantages. One such machine, the Cybex II, uses a hydraulic and electrical system to maintain a constant speed through an exercise, no matter how hard a person pushes. Indeed, the harder you push, the harder the machine pushes back. Sports demand not only strength but strength at specific speeds. By approximating an athlete's speed of movement, isokinetic machines theoretically offer more efficient training. Cybex machines were originally used in hospitals and clinics to help patients regain their strength. The Cybex II includes a monitor and printed record resembling an electrocardiogram that enables physicians and therapists to see precisely how much force a muscle exerts at any point in an exercise motion.

Anatomy and Destiny

All athletes are blessed or cursed with some characteristics that no barbells, regimen or drug can change. To some extent, great athletes are born, not made. The best thing a hopeful superstar can do is choose his parents wisely.

At the cellular level, the mixture of slow- and fast-twitch muscle fibers clearly influences who excels at what sports. Slow-twitch fibers, with their greater aerobic capacity, are best suited for endurance sports. Fast-twitch fibers, with a higher anaerobic capacity and a shorter contracting time, predispose a muscle toward power and speed. One kind of fiber cannot be converted into another, so in this realm, anatomy would seem to be destiny. Muscle biopsies, which entail removing bits of tissue from various muscles, have revealed that most long-distance runners have a high percentage of slow-twitch fibers. In 1975, David Costill of Ball State University took muscle biopsies of some of the best distance runners in the U. S. and found that the average was 79 percent slow-twitch and 21 percent fast-twitch. Similar tests on shot putters, discus throwers and sprinters show a different average mix of fibers — 27 percent slow- and 63 percent fast-twitch. But there are exceptions. In Costill's study, one long-distance runner had 98 percent slow-twitch fibers, physiologically ideal for a marathoner. But another runner, also an excellent long-distance man, had only 53 percent slow-twitch fibers. Such tests suggest that the mix of muscle fibers may be a good general indicator of an athlete's best sport, without being an immutable law of sports. Many athletes, especially the best ones, are almost obsessively competitive,

Cybex II machines build strength through isokinetic contractions. The machine controls the speed at which athletes complete a motion. Adjusting the machine enables an athlete to build strength at different speeds. Originally used in rehabilitative therapy, Cybex II also records the force an athlete or patient exerts, pinpointing areas of weakness that linger after an injury or illness.

willing to drive their bodies further and harder than most other people. Such strength of will can push muscles of any make-up a long way.

Many other factors besides the percentage of fiber types in muscle influence athletic ability. A six-foot-two-inch man or woman has a decided edge in basketball over people of average size. Gymnasts, as a rule, are fairly short, which may give them a lower center of gravity and better balance. Various sports tend to attract participants of certain sizes and shapes, at least at the higher levels of competition, and the process seems to reinforce itself. A small, lean teen-ager may find himself better than his friends at running long distances. His slight edge in natural ability could well persuade him to pursue the sport. If he trains as a long-distance runner, building his ability to draw oxygen and overloading his slow-twitch fibers, then his innate superiority increases. Thus, hypothetically, are marathoners born and made.

Temperament, aggressiveness, coaching and opportunity also play a role. But studies have shown that people of a certain build excel at certain sports. Discus throwers tend to be large and strong, pole vaulters shorter and more muscular, high jumpers tall and slender. English doctor J. M. Tanner analyzed 137 athletes at the 1960 Rome Olympics and found a consistent pattern of similarities among athletes in the same events. Sprinters had short legs, large limbs and were very muscular in general. Hurdlers tended to look like larger, long-legged sprinters. Middle-distance runners were larger than sprinters with longer legs and broader shoulders. Short legs, narrow shoulders, light musculature and small stature characterized marathoners. Weightlifters were powerful and heavily muscled but short in stature and limb. Discus throwers were the largest athletes of the group, while high jumpers had the longest legs in relation to their torsos.

In broader terms, people can be classified by physique under a system called somatotyping. Different approaches to somatotyping exist, but the most widely known is psychologist William H. Sheldon's technique, which groups people by their fatness, muscularity and leanness — in Sheldon's terms, endomorphy, mesomorphy and

Exploding off the starting platform, swimmers share a moment of flight. Top athletes in any sport show similar physiques. Broad shoulders and slender bodies mark top swimmers around the world.

ectomorphy. Endomorphs are round, heavy people characterized by large internal organs and an appearance of fatness and softness. Mesomorphs are strong and heavily muscled. Ectomorphs tend to have slender limbs and a light musculature, in rough terms to be light and slight.

With Sheldon's system, every physique can be assigned a numerical tag that describes its degree of endomorphy, mesomorphy and ectomorphy. A man whose somatotype is 7-1-1 would be heavy, perhaps obese, narrow-shouldered and large-hipped. A somatotype of 1-7-1 describes the archetypal mesomorph, broad-shouldered, solid and muscular. One scientist uses the word "beanpole" to describe ectomorphs. Numerically, the somatotype of an exaggerated ectomorph would be 1-1-7. Few people lie at the extremes of any of the three somatotypes. Most people have somatotypes like 5-3-2, a little on the heavy side, or 3-5-2, an athletic build. A somatotype of 4-4-4 might describe the elusive human that so many surveys, studies and opinions depend upon, the fabled average man.

Athletes tend somewhat toward mesomorphy and ectomorphy, although this rule depends on the sport. Marathoners may be the best example of ectomorphs in sports, although even they are not pure ectomorphs. Mesomorphy calls to mind sprinters, football players and wrestlers or gymnasts. The endomorphic combination of excess fat, narrow shoulders and broad hips work against the agility, balance and upper body strength needed in sports. There are exceptions, however. Chris Taylor, an American superheavyweight wrestler, advanced to the finals of the 1972 Olympics in Munich. Taylor weighed 440 pounds and obviously showed endomorphic characteristics, but used size to his advantage. Although mesomorphs have a physique that equips them well for most kinds of athletics, not all mesomorphs are athletes. The world's finest endurance athletes, as well as many high jumpers and basketball players, show signs of ectomorphy. And as Chris Taylor proved, endomorphs can excel in sports.

Two other major distinctions among human beings influence athletic performance — race and sex. Race is a subject many scientists shy away

A portly and a thin man, endomorph and ectomorph, stand at poolside in Business Men's Bath *by George Bellows. Endomorphy and ectomorphy, roughly fatness and leanness, are characteristics used by psychologist William H. Sheldon to classify human physiques, a system called somatotyping.*

from. Even careful, open-minded studies of the subject are liable to misinterpretation, and sometimes willful distortion. General observations about race and athletic performance are easy to come by, but of dubious value. The Japanese men's gymnastics team dominates its sport, and has since the 1960s. Great height is a disadvantage in events that demand near perfect balance and coordination. Japanese men tend to be short. It is a characteristic of their race, and one that may give them an advantage in gymnastics. But countless other factors make the connection between race and athletics tenuous. The Japanese men's team enjoys a tradition of fine gymnastics. Excellence breeds excellence. As a nation, Japan may encourage gymnastics and discourage other sports. Japanese teams may have the world's best gymnastics coaches. Teasing these factors apart to determine which ones are responsible for athletic excellence is an impossible task.

But difficult or not, the issue shows no signs of disappearing. In the United States where roughly 12 percent of the population is black, 65 percent of the professional basketball players, 42 percent of professional football players and 19 percent of professional baseball players are black. In track and field, as of early 1982, black men held the world records in the long jump, the 100-meter and 400-meter runs and both the 110-meter and 400-meter hurdles. The achievements of black athletes would seem to suggest that in the sports demanding speed, agility and jumping ability, blacks excel.

Many scientists believe that the high percentage of blacks in professional sports reveals more about American culture than human physiology. Many black professional athletes are well paid and glamorous, making them appealing models for black children. Sociologist Harry Edwards of the University of California at Berkeley believes that disproportionately high numbers of black children are drawn or pushed toward athletic careers, where success will necessarily bless only a few. In American society, black presidents, senators and heads of corporations are less common. Their comparative lack of visibility makes them less accessible as models and reinforces the stature of athletes. Some evidence suggests that race is a physiological factor in sports, but there is no proof. It is surely a cultural factor. Jesse Owens, a black man, a winner of four Olympic gold medals and probably the finest athlete of his day, said: "There is no difference between the races. If the black athlete has been better than his white counterpart, it's because he's hungrier — he wants it more."

J. M. Tanner's study of the physiques of athletes at the 1960 Rome Olympics touched on physiological differences between white and black athletes. In all the events he examined, blacks had longer arms and legs, smaller calf muscles and narrower hips than their white counterparts. "From a mechanical point of view," Tanner wrote, "a Negro with identical legs to a white would have a lighter, shorter, slimmer body to drive, apart from the arms."

Within the broad categories most people use to distinguish among the races — white, black and

A Japanese gymnast personifies strength, control, grace and mesomorphy. The first three attributes are hallmarks of his sport, the last of his physique. Mesomorphy, or muscularity, characterizes most athletes, although to some degree, everyone is a mixture of fatness, muscularity and leanness.

Jesse Owens, many believe, was the greatest athlete of his day. Hitler hoped the 1936 Berlin Olympics would glorify the Aryan race. But the hearts of the crowd and four gold medals went to Owens.

yellow — subcategories exist that further refine the connection between race and athletic ability. The term Negro, used in its strictest sense, refers to blacks of southern, western and central Africa. Black and dark-skinned people of eastern and northern Africa are classified in some schemes as Hamites. According to the theory, great black distance runners like Kipchoge Keino and Henry Rono, who today holds the world record in the 3,000-, 5,000- and 10,000-meter runs and the 3,000-meter steeplechase, are Hamites. Racial advantages supposedly give Negroes an advantage in sprints and endow Hamites with a physiological edge in long-distance events.

Physiological differences among races and even smaller ethnological categories may exist, but as centuries pass and races mingle, the distinctions necessarily blur. And statistics can say many things. No black women hold world records in sprints, hurdles, relays or jumps. If blacks possess a racial superiority in these events, world records would seem to suggest that it applies only to men. Since 1960, no black man has won an Olympic gold medal in high jumping. As far as racial advantages go, jumping and high jumping would seem to be two different things. The physiological influence of race on athletic ability is only a question, and certainly no predictor of success in sports or any other field.

Sex is another matter. Men and women are built differently, a discrepancy that gives an advantage to men in most sports. The physiological differences are numerous, although none bear heavily on athletic ability until puberty. Before puberty, boys and girls are roughly the same size and strength. If either sex has an athletic advantage, it is the female. Puberty occurs in females about two years before males. It is an important difference, because the onset of puberty slows bone growth. By maturing later, males have a few more years to grow. Girls also begin to add fat at puberty, boys, muscle. By the late teens and early twenties, about 25 percent of a woman's weight is adipose tissue, or fat, compared to 14 percent for men. In athletes of both sexes, the percentage of body fat can be considerably lower, but the general difference between men and women remains. One study found that male gymnasts carried about 6 percent of their weight in fat, on the average. For female gymnasts, the figure was 16 percent. One male marathoner had only 1 percent body fat, a figure physicians would once have considered dangerous. For a female marathoner, 6 percent was the lowest figure. Male swimmers in the study averaged 9 percent body fat, females, 24 percent.

Larger Hearts, Thicker Blood

The heart, lungs and bones of males are larger than those of females. Men have a slower heart rate, which, coupled with their larger hearts, enables them to pump more blood with fewer beats. The blood of a typical male is thicker than a woman's. It carries about 6 percent more red blood cells per cubic millimeter, enabling men to draw slightly more oxygen into their blood streams. Men also have more and bigger muscle fibers than women. In addition, the male hormonal system produces testosterone, the chemical on which anabolic steroids are modeled. One of the hormone's effects is to increase muscle size. Even the muscles of women who lift weights do not hypertrophy, or enlarge, to the extent that men's do because their bodies lack testosterone. Women should not avoid strenuous

Blacks dominate professional basketball and other sports. Although some observers suspect a physiological factor, others see subtle cultural currents that draw blacks into sports and away from other fields.

athletics and even regular strength training for fear of building bulging muscles. Their muscles grow stronger, just as men's do, but gains in size are much less pronounced. A woman's heart rate, cardiovascular capacity, strength and endurance also improve in the same way a man's do. She stands to gain as much as a man from exercise, physiologically and psychologically.

Women have won Olympic medals during all phases of their menstrual cycles. Nor, apparently, does motherhood preclude excellence in athletics. Fanny Blankers-Koen of the Netherlands was, at thirty, the oldest female athlete at the 1948 Olympics. In a week of competition, Blankers-Koen, a mother of two, ran eleven races and won four gold medals. In the 1976 Olympics, Poland's Irena Szewinska, the mother of a six-year-old, set what was then a world record in the 400-meter run. A study of Hungarian women showed that female athletes had shorter labor and fewer caesarian sections than their less athletic peers.

A Better Balance

In a few areas, the athletic abilities of women exceed those of men. Their wider hips give them greater flexibility in their lower bodies, a lower center of gravity and better balance. The slightly higher percentage of fat in the body probably helps women in endurance events, enabling them to burn more fat and less glycogen, and thus stave off exhaustion. Female endurance swimmers benefit doubly from their fat. It is a source of energy and buoyancy, because fat floats. The only athletic event in which women have consistently outperformed men is the English Channel swim. In 1926, Gertrude Ederle swam across the Channel in 14 hours and 31 minutes, two hours faster than any man had done it before. Another American woman, Penny Dean, holds the record today, at 7 hours and 40 minutes.

Pound for pound, a woman's muscle is as strong and steady as a man's. The athletic superiority of men is a matter of more, not better, muscle. Even in a sport many people would consider a male preserve, powerlifting, some women excel. Thirty-year-old Jan Todd of Alabama, is one of the world's strongest women. Powerlifting consists of three events: the deadlift, crouching to

Babe Didrikson follows through at the tee. Writer Grantland Rice called her the most flawless example of "muscle harmony, of complete mental and physical coordination the world of sport has ever seen."

Billie Jean King bends to return a low volley. King was ranked first in the world in her sport four times between 1965 and 1973. The stature of women's tennis today owes much to her efforts and talents.

lift a weight from the ground and then rising to a standing position; the bench press, lying on a bench with arms extended, lowering a weight from that position to the chest and then raising it again; and the squat, lifting a weight off a rack, doing one knee bend and raising it again. In the super-heavyweight class, Todd holds world records at 545 pounds in the squat and 1,230 overall.

The history of sports is filled with outstanding female athletes. Mildred "Babe" Didrikson excelled at every sport she tried, track and field, baseball, basketball, golf and others. She won two gold medals, in the 80-meter hurdles and the javelin throw, in the 1932 Los Angeles Olympics. She won the Associated Press Woman Athlete of the Year Award in 1932, 1945, 1946, 1947, 1950 and 1954. Wilma Rudolph was afflicted with double pneumonia and scarlet fever as a child, which left her unable to walk until she was eight years old. In the 1960 Olympics, at age twenty, she won three gold medals, the 100-meter and 200-meter runs and the 400-meter relay. Tennis star Billie Jean King holds more Wimbledon championships than any player, male or female.

Women athletes are also improving in comparison to men. In the 1950s, women's times in the 800-meter run averaged 19.4 percent slower than men's. But by the 1970s, their times were only 11.2 percent slower. As more women enter sports, the gap will narrow further. Perhaps only in long-distance swimming events, however, do women stand a good chance of surpassing men. The greatest impediment to women's athletics is a lack of encouragement and support, financial and cultural.

As women athletes cut seconds off their times in races, they will be following a tradition that has held true since modern Olympics began eighty-six years ago. No Olympic record from fifty years ago stands today. Ernst Jokl, professor emeritus at the University of Kentucky, sees several reasons for the steady, sometimes dramatic improvement in athletic performances over the course of the twentieth century. Better health, nutrition and medicine have enabled more people to live longer. More people constitute a larger pool from which to draw athletic talent. Technological innovations, such as the fiber glass pole for pole vaulters, have radically changed some sports. Identifying gifted athletes early has also helped bring the most talented men and women into competition. Finally, Jokl declares, sports, like other fields, will always be blessed with a few geniuses, people who display an unpredictable brilliance at their sport.

Remarkable athletes turn up in every generation. Powerlifting may have one in nine-year-old Tammy Stafford of Albuquerque, New Mexico. In the deadlift, Tammy, all forty-nine pounds of her, can raise a 250-pound barbell from the floor to above her knees. A senior in high school in Riverside, California, six-foot-two-inch Cheryl Miller can dunk a basketball. She is perhaps the only woman ever to do so in organized competition. Miller once scored 105 points in a single game. In the summer of 1981, English runner Sebastian Coe set world records in three of the most competitive of track and field events, the

Tammy Stafford, age nine, strains to lift 450 pounds in the hack squat. Girls and women were discouraged for centuries from participating in many sports. Their athletic abilities remain largely unexplored.

800-meter, 1,000-meter and mile runs. Coe has posted the current world record for the mile at three minutes, forty-seven and thirty-three one-hundredths seconds.

Perhaps the most extraordinary effort in sports history occurred at the 1968 Olympics in Mexico City. American long jumper Bob Beamon sprinted toward the takeoff board, planted his foot and leaped into the air, pumping his legs for more distance. Beamon landed twenty-nine feet, two-and-a-half inches away. His jump broke the old record by two feet. No other athlete has come within a foot of his incredible performance since. Some scientists suspect that if Beamon had run any faster or hurled himself into space with any more power, the strain of his jump might have torn the muscles of his leg from their attachments at the hip.

Jokl calls Beamon's jump a "mutation performance," possibly the best the human organism can do in that event. Contemporary athletes are still improving on earlier records, but progress is coming harder, at least in men's track and field competition. By the year 2000, Jokl suggests, sprinters may cut two-tenths of a second off of American Jim Hines's mark of 9.95 seconds for the 100-meter run, but such an achievement is by no means assured. Hines improved on the 1960 Olympic record by three-tenths of a second, and no one has surpassed his mark for fourteen years. Over the next twenty years, the biggest improvement will almost certainly come in women's sports. Top women athletes have equaled or surpassed the 1912 records of men in the 100-meter run, the marathon, the high jump, the 400-meter relay and other sports. In 1980, at Moscow, Ines Diers of East Germany bested American swimmer Don Schollander's 1964 Olympic gold medal performance in the 400-meter freestyle. Greta Waitz's 1979 time in the New York marathon — two hours, twenty-seven minutes and thirty-three seconds — would have won the event for both men and women in 1970. But the human body has athletic limits, which Bob Beamon may already have attained in the long jump. Other sports await the arrival of other geniuses, from whom no record is ever completely safe.

Crumbling Dreams

The pursuit of athletic excellence, whether one achieves it or not, exacts a price. The greatest athletes pay with hours of training and sacrifice, and some of them are rewarded with money, fame and accomplishment. The less great pay in disappointment, disillusionment and sometimes even more valuable coin. From elementary school to the pros, more than bones are broken when an athlete falls to injury. Too often, dreams crumble as well. The esteem America reserves for its athletes cannot help but influence even the youngest participants, encouraging some to compete with a ferocity befitting warriors rather than Little Leaguers. Injuries among small children in sports are thankfully rare. Small bodies and immature muscles do not generate the same irresistible forces that endanger life and limb in professional sports. But because children's joints and bones are not fully formed, they can suffer injuries that will plague them all their lives. In a child, the ends of arm and leg bones are equipped with a growth plate, a layer of tissue from which new bone grows. Fracture the growth plate and the bone may grow irregularly, leaving one leg shorter than the other or one arm shaped differently from its partner.

As children grow stronger and the competition becomes more fierce, injuries increase proportionately. Studies of high school football players show that 40 to 50 percent suffer injuries yearly. A group of high schools in the area of Seattle, Washington showed an 81 percent injury rate. Every year roughly 250,000 high school football players are injured seriously enough to miss at least a week of practice. The number of teenagers playing high school football is so great that sports physician James Garrick says, "More high school kids get injured every Friday night than pros do in a year." This is a chilling accomplishment when one considers that the injury rate in the National Football League hovers around 100 percent. In professional football, there are at least as many injuries as people.

When Sport Turns Ugly

Often, the violence that simmers beneath the surface of many sports erupts into open brutality. Professional hockey player Teddy Green of the Boston Bruins played his last two seasons with a metal plate inserted in his head after his skull was fractured in a hockey brawl. The most dangerous outbursts of violence involve fans. In 1964, more than 300 people in Lima, Peru were killed in a fight and stampede that erupted after a soccer match. Soccer can even lay some claim to starting a war. In 1969, after two bitter soccer matches, El Salvador invaded Honduras. The two countries needed no sports to encourage their mutual hostility. But the rabid nationalism that burned at their soccer matches served as a convenient pretext for their short war.

The allure of professional sports destroys athletes in other ways. Thousands of outstanding high school athletes go to college every year on athletic scholarships. Some do not graduate from high school first. Some are illiterate, but graduate anyway. Some buy specious credits or doctored

Bob Beamon, leaping almost into flight, sets the world record in the long jump at the 1968 Mexico City Olympics. Many experts consider this jump the most extraordinary athletic feat of all time.

transcripts from dishonest officials at junior colleges, although certainly most athletes who win athletic scholarships leave high school more or less under their own steam. But too often their athletic ability carries them through four years of college sports and no further. Only 2 percent of college athletes make it to the pros. Some colleges struggle mightily to keep athletes academically eligible for four seasons of football or basketball, only to abandon them during the last semester or two of college. At that point, some scholarship athletes face a formidable list of required courses that they must pass to graduate.

Many simply do not make it. Edwards estimates that 60 to 70 percent of the black athletes who go to college on football and basketball scholarships do not graduate. According to the National Football League Players Association, two-thirds of the players in the NFL do not have college degrees. In 1978, several former athletes

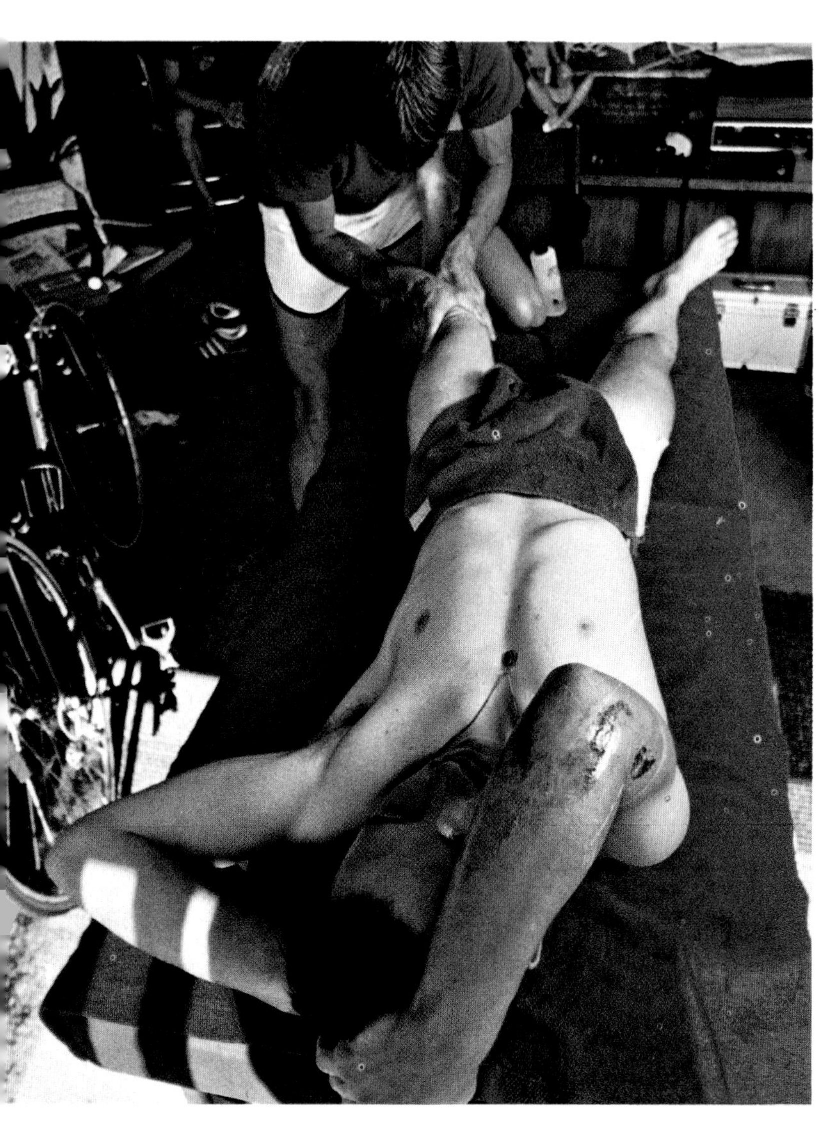

Irish bicyclist Alan McCormack, resting after a crash, allows a trainer to knead his aching muscles while he holds a battered arm over his head. At the highest levels of competition, many sports, even bicycling, court danger.

from California State University at Los Angeles brought suit against the university for breach of contract. Their scholarships, the athletes argue, implied an agreement by the university to help educate them. Cal State therefore failed to honor this agreement. The case is still pending.

The responsibility for so many athletes failing to graduate rests with both the athletes and colleges. At many colleges and universities, officials believe their responsibilities are fulfilled if they merely keep an athlete in school for four years. It is the athlete's responsibility, some educators feel, to take advantage of his opportunity to get an education. But too many football and basketball players refuse to accept that they are good but not great, convinced that a career in the pros awaits them. Laboring under the illusion that their futures are secure, they coast through easy courses and never achieve a balance between the demands of their sport and the rigors of academics. When the truth becomes plain — that they just are not good enough for the pros — they are unprepared for any other profession.

The Dark Underside

Money drives this system of athletic success and academic failure. During the 1982-83 and 1983-84 seasons, two major television networks, CBS and NBC, will pay the National Collegiate Athletic Association (NCAA) $263 million for the right to broadcast college football games. CBS will pay another $48 million to air college basketball games. Big time athletic departments have big budgets. In 1980, James Odenkirk, head of the physical education department at Arizona State University, surveyed the athletic budgets of several major universities and found the following: Ohio State, $9.7 million; University of Michigan, $8 million; University of Oklahoma, $5.5 million; UCLA, $5.25 million. The figures speak eloquently: college sports are big business.

All this money generates tremendous potential for fraud and scandal. And the millions of dollars gambled on sports every year only make matters worse. Professional baseball was rocked by the Black Sox scandal in 1919. For $80,000 divided unequally among them, eight players from the favored Chicago White Sox conspired to lose the

Sports reflect personality and culture. A boxing match can be a simple exploration of strength between boys or a destructive fight for honor and esteem between unready and unwilling warriors.

World Series. One player, Arnold Gandil, took $35,000 for his work as ringleader. No one knows how much the 1919 World Series made for gamblers, who bet close to a quarter of a million dollars on the fix.

College sports are hardly immune. Over the last thirty years, players from the City College of New York, Long Island University, New York University, the University of Kentucky, Boston College and several other schools have been convicted of point shaving. In point shaving, players conspire not to win or lose by too much, enabling gamblers to win bets on the margin of victory. The irony of the practice is that some players who have not graduated from high school, who play football or basketball because they have phony transcripts, are expected to hold the outcome of the game sacred. Some do not.

Professional sports shine brightly on some athletes. Earvin "Magic" Johnson of the Los Angeles Lakers basketball team will earn $25 million over the next twenty-five years. But the astounding popularity and wealth of athletic superstars cast a shadow over thousands of other aspiring athletes who cannot be as good and cannot bring themselves to believe it. The dream drives children, with and without their parents' blessing, into football, baseball, hockey and boxing before they finish elementary school. For the luckiest and most talented, the dream carries them to an athletic career and an enviable income, if it does not abandon them somewhere along the way with a useless knee or a pain that will never go away. Others fortunately discover that the dream is not for them, and begin looking for another. The victims are the athletes who cannot use their talents to earn a living and have not learned to do anything else. The dark underside of the dream, something many athletes realize too late, is that the Magic Johnsons are so rare.

Appendix
Limbering Up

About 60 percent of the sport-related injuries seen at the Sports Medicine Resource center in Massachusetts result from improper training. The point at which judgment and not sore muscles should be exercised depends on physical conditioning, age, sex and weight. Generally, any of the following signs signal impending injury if precautions are not taken:

1. Localized pain or stiffness in a joint accompanied by loss of motion.
2. Swelling.
3. Burning, tingling and numbness in a limb.
4. Unexplained redness or black-and-blue marks.
5. Change of contour in bone, joint or muscle.

Ignoring these signals can exacerbate the problem. If pain or warning signals persist, consult a doctor.

If you are injured from pulling, spraining or straining a part of your body, discontinue exercise and other activities.

Elevate the injured part to a height above the level of the heart so that any excess fluid that could cause swelling will drain. Wrap an ice pack firmly against the injured part of the body with an elastic bandage. This, too, helps control swelling by shrinking torn blood vessels. Consult a doctor.

One of the best ways to prevent an injury in the first place is to begin a regimen of stretching exercises followed by a suitable warm-up before

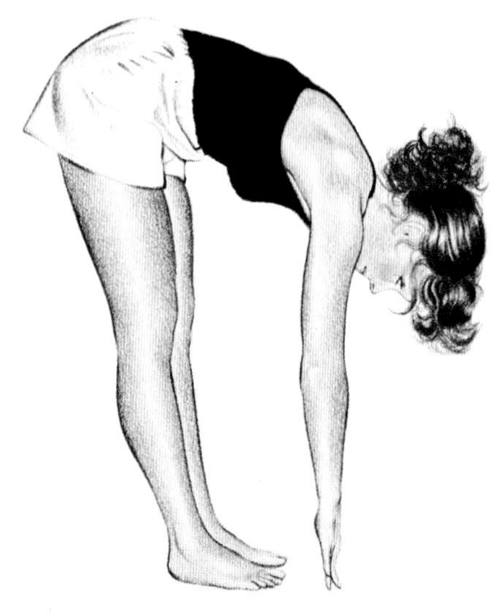

participating in sports. Each time you subject your body to stress in vigorous competition, you injure your muscles. When they heal, the muscles tend to shorten slightly. Stretching lengthens muscles and tendons, filling them with blood that is rich with vital nutrients. Because stretching tends to make muscles more pliant, they are less likely to sprain or to strain.

Stretching should be done deliberately and without bouncing. It should feel comfortable and relaxing. Stretch until reaching an easy tension, then hold position and relax, concentrating on the feeling of the stretch. As the sensation decreases, stretch a little further until achieving the same awareness of a comfortable stretch. Effective stretches can last sixty seconds and should involve no pain. Muscles become shorter with aging. The older one gets, the more advisable stretching becomes.

These are five basic stretching exercises for running sports that constitute a good start to a warm-up regimen. Rest briefly between sets. Do only as many as is comfortable when starting. People who are already in good physical condition should stretch for at least fifteen minutes before vigorous exercise or competition.

Floor touching:
Stand straight with heels together in a relaxed stance. Keeping knees straight, try to touch your fingertips to the

floor without bouncing. Hold the position for ten seconds and release. Repeat this exercise at least five times.

The plow:
Lie on your back and raise your legs overhead, keeping knees straight. Without forcing, try to touch the floor with your toes. Hold position for ten seconds, then lower your legs. Repeat five times.

The Japanese split:
Standing with your knees straight, gradually spread the legs apart as far as possible and place your palms on the floor for balance. Hold this position to the count of ten. This stretch should be done at least five times.

Wall push-up:
Place the palms of your hands on a wall about four feet away while standing in a relaxed stance. Bend your elbows so that your upper body moves closer to the wall. Keeping heels flat on the ground, the calves and Achilles' tendons will stretch. Hold this position for ten seconds. Straighten your elbows and repeat at least five times.

Stair stretch:
Standing with your feet halfway off the edge of a step or curb, rise up on your toes. Hold this position to the count of ten. Next, lower your heels below the step, and hold this position for ten seconds. You will feel the stretch in your calf muscles and Achilles' tendons.

After completing the stretching exercises, begin the warm-up to raise muscle temperatures a few degrees. This helps them to perform more efficiently under the stress of sport. The heart and lungs, too, will pump more vigorously in anticipation of more strenuous demands placed upon the body during actual play.

When you warm up before participating in a sport, you should exercise your muscles in the same ways as you will use them in actual athletics. In the following weeks, gradually increase the intensity of warm-up exercise at a pace that muscles can easily adjust to.

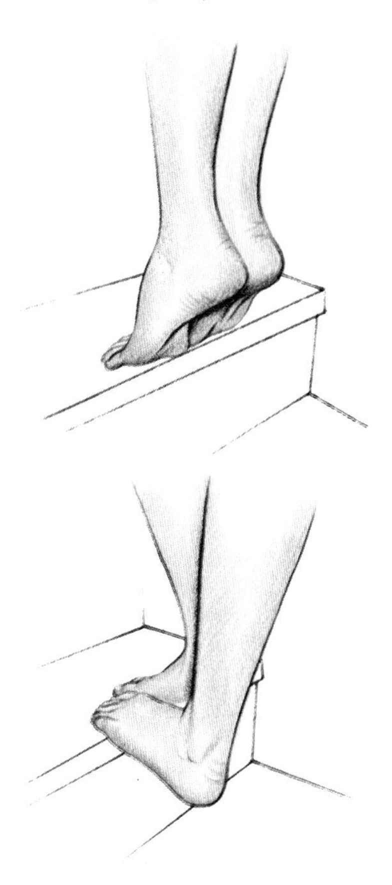

Glossary

abductor any muscle that draws parts of the body away from the central trunk.

abductor hallucis the muscle that flexes the great toe.

actin one of the two main contractile proteins in a muscle fiber that forms the thin filaments.

adductor a muscle that draws body parts toward the trunk.

adductor magnus the large muscle situated on the inner thigh that extends the hip joint.

adenosine triphosphate (ATP) a chemical compound that breaks down to release the energy responsible for muscle contraction.

aerobic with oxygen, or in the presence of oxygen.

"all or none" principle the law of muscle contraction holding that muscle fibers in a motor unit contract maximally or they do not contract at all.

American Sign Language the leading form of manual communication among deaf people in the United States.

amyotrophic lateral sclerosis "Lou Gehrig's disease;" a progressive, paralyzing disease marked by the destruction of motor neurons in the brain or the spinal cord.

anabolic steroids synthetic derivatives of the male hormone testosterone that promote tissue growth.

anaerobic without oxygen, or in the absence of oxygen.

anaerobic glycolysis the breakdown of sugars in the absence of oxygen; one of the major methods of producing energy in muscle fibers.

antagonist a muscle whose contraction produces a joint action opposite to that of the prime mover.

assistant movers the muscles that contribute to a specific movement, assisting the prime mover.

ATP adenosine triphosphate.

barong a religious dance of Bali marked by expressions of fear to appease malevolent forces.

basal ganglia a group of nerve clusters deep in the interior of the cerebrum which are thought to be involved in the initiation of voluntary movements.

biceps brachii the prominent fusiform muscle on the front of the upper arm.

biomechanics the study of the mechanical forces, such as drag, torque and inertia, at work in the movement of living bodies.

biopsy the removal and examination of tissues from the living body.

blood doping the removal and later reinfusion of blood in the same person for the purpose of increasing the number of red blood cells and blood volume.

body language movements and gestures of the body that may indicate one's true feelings.

botulin a lethal bacterial toxin that causes paralysis by blocking the release of acetylcholine at the neuromuscular junction.

buccinator a facial muscle that compresses the cheek and draws back the mouth.

cardiac muscle one of the body's three types of muscle, found only in the heart.

center of gravity the imaginary point in the body where weight is equally distributed.

cerebellum an oval mass behind the cerebrum which coordinates automatic, voluntary muscle movements.

cerebral cortex the furrowed outer layer of the brain where higher mental functions reside.

chorea a group of neuromuscular diseases causing jerky, spasmodic, involuntary movements.

choreography the arrangement of the movements and steps of a dance.

connective tissue the supporting and connecting structures of the body.

contraction the development of tension within muscle.

creatine phosphate a molecule in muscle fibers that breaks down to liberate energy for the construction of adenosine triphosphate (ATP).

cross bridges projections of myosin molecules that form links with actin filaments and pull them forward, causing contraction.

cuirasse esthétique a style of Greek sculpture in which the musculature of the torso was clearly delineated as it joined the top of the legs.

curare a resin extracted from several tree species which impedes neuromuscular transmission.

dance notation a system of recording the movements of dance on paper.

deltoid muscle the triangular muscle of the shoulder that is active in all shoulder movements.

diaphragm a muscular membranous partition that separates the abdominal and thoracic cavities.

dithyramb an impassioned dance performed by men in Dionysiac rites of ancient Greece.

dorsiflexion backward bending.

dorsiflexors muscles that bend body parts back.

Duchenne muscular dystrophy the most common form of muscular dystrophy, which usually attacks young boys.

ectomorph a thin person; an individual with a predominantly lean physique and light musculature.

electromyograph (EMG) an instrument that detects and records electrical activity in muscles.

electron a negatively charged subatomic particle.

electron microscope a microscope that uses electrons rather than visible light to produce magnified images. Extremely powerful, it is usually used to create images of objects smaller than the wavelengths of visible light.

en pointe a technique of ballet in which a ballerina executes a variety of steps on her toes.

endomorph a heavyset person; one whose physique is characterized mainly by roundness and softness.

endomysium a sheath of connective tissue that surrounds each muscle fiber.

endorphin a brain chemical that eases or suppresses pain.

enzyme a protein that promotes the chemical processes occurring in living tissue without itself being altered or destroyed.

esophagus a muscular tube that carries food to the stomach.

extensor muscles muscles that straighten body parts away from the main body.

external perimysium connective tissue that links bundles of muscle fibers together to form muscle.

extrapyramidal system a major motor transmission network consisting of a variety of nerve relays in the brain and spinal cord; thought to control large, automatic body movements.

fasciculus a group of muscle fibers.

fast-twitch fiber one of the two major types of skeletal muscle fibers that contracts and fatigues quickly and relies mainly on anaerobic energy.

fatty acid one of the building blocks of fats; used as fuel for muscle contraction.

fibril a small fiber; see myofibril.

first-class lever system the arrangement of bone and muscle in which the fulcrum lies between the force and the resistance.

first law of motion the law in physics stating that an object will remain at rest or in motion unless it is compelled to change its state by external forces.

fixator a muscle that holds a bone or body part steady, providing a firm foundation for muscles to pull against; a stabilizer muscle.

flexion the act of bending a body part.

flexor digitorum brevis a short muscle that flexes the toes.

folk dance a traditional dance originating among the common people of a nation or region, often embodying religious beliefs and other customs.

fox trot a rapid dance with a running step that originated in the United States around the time of World War I.

fulcrum the point on which a lever turns.

fusiform tapered at each end; spindle-shaped.

galliard a social dance popular in England in the sixteenth century.

glucose the most common sugar and the main fuel for muscle contraction.

glucose sparing the preference of muscle fibers for fatty acids as fuel rather than glycogen or glucose.

gluteus maximus the large muscle in the human body which extends, abducts and rotates the thigh to the side.

glycogen a carbohydrate of several glucose molecules; the main form of glucose storage in the body.

glycogen loading following a regimen of exercise and diet to increase stores of glycogen in muscles.

glycolysis the breakdown of glucose, or sugar.

Golgi tendon organ a sensory receptor located in tendons, near their junction with muscles, that relays information about muscle tension.

grand jeté in classical ballet, a high, athletic leap, with one leg extended in front of the body and one behind.

grief muscles a term used by Charles Darwin to denote the muscles of the forehead that are habitually employed to express sadness.

growth hormone a substance that stimulates growth.

growth plate a thin layer of tissue near the end of the shaft of a long bone that enables the bone to grow in length; also called epiphyseal plate.

gymnastai professional athletic coaches and trainers in ancient Greece.

hamstring a fleshy tendinous muscle that courses from the hip to the rear of the knee.

hypothalamus a tiny mass of nerves below the thalamus that regulates basic body functions, houses centers for pain and pleasure and influences movements associated with emotions.

inertia the tendency of an object to resist acceleration.

intercalated disks strong binding sites between individual cardiac muscle fibers. Their low electrical resistance allows them to pass commands quickly to contract.

internal perimysium the connective tissue that surrounds groups of muscle fibers, the fasciculi.

interneuron a neuron in the brain or spinal cord that conveys impulses between a motor neuron and a sensory neuron.

isokinetic the kind of muscular contraction in which muscles contract at a constant speed against varying degrees of resistance.

isometric the kind of muscular contraction in which muscles contract against immovable resistance and do not change length.

isotonic the kind of muscular contraction in which muscles contract at varying speeds against a constant resistance.

jitterbug an athletic dance popular just before World War II.

Kabuki a traditional dramatic dance of Japan that relates stories of moral intrigue through stylized movements and elaborate dress.

kinesiology the study of motion of the human body.

kinesthesia the sensations of body position and movement derived from sensory organs.

kordax a bawdy, masked dance performed in theatrical productions of ancient Greece.

Labanotation the method of dance notation developed by Rudolf Laban.

lactic acid a product of glucose and glycogen metabolism, which, in sufficient concentration, causes fatigue.

latissimus dorsi the broad muscle located at the lower half of the back which is a prime mover for adduction, extension and hyperextension of the shoulder joint.

laws of motion principles developed by seventeenth-century English scientist Sir Isaac Newton to explain the physics of movement.

leupeptin a chemical that inhibits the action of proteinases.

levator scapulae the small muscle situated on the back and side of the neck that is the prime mover for shoulder girdle elevation.

lever systems the arrangement of levers in the body that permits muscles to perform work by exerting tension on the points of their insertion into bones.

ligament a band of fibrous tissue that supports and strengthens a joint by connecting bones or cartilage.

load arm the zone between the load to be managed and the fulcrum.

lockjaw one of the first signs of tetanus, in which the victim's jaw is firmly clenched against his will.

lordosis a postural defect caused by habitually assuming a swayback stance.

medulla a portion of the brainstem, continuous with the spinal cord, serving as a crossroads for motor pathways.

mesomorph a person whose physique is marked predominantly by powerful musculature.

minuet a refined dance with delicate steps, popular throughout Europe in the seventeenth and eighteenth centuries.

mitochondrion one of many small sacs inside muscle fibers and other cells containing enzymes that help convert food to energy.

momentum the force of motion; the product of a body's mass and linear velocity.

mononucleate having a single nucleus.

motor homunculus a schematic drawing showing the relative neuronal representation of the body's muscles in the primary motor cortex.

motor neuron a specialized cell of the nervous system that transmits impulses from the brain or spinal cord to a muscle or gland.

motor unit the basic unit of movement consisting of a motor nerve fiber and all of the muscle fibers it supplies.

multinucleate having several nuclei.

multiunit smooth muscle one of two types of smooth muscle, made of numerous loosely knit, independent fibers controlled largely by nerves.

muscle fiber a muscle cell.

muscle spindle a sensory receptor that detects changes in muscle stretch.

muscle tone the degree of vigor and tension in muscle.

muscular dystrophy a general term for a group of nine diseases causing irreversible wasting of muscle.

myasthenia gravis a neuromuscular disorder characterized by extreme muscular weakness, fatigue or paralysis caused by the inhibition of nerve transmission at neuromuscular junctions.

myoblasts small, mononucleate muscle cells that merge before birth to form larger, multinucleate skeletal muscle fibers.

myofibril one of many slender fibrils, consisting mainly of protein, that fill a muscle fiber.

myoglobin an oxygen-transporting protein in muscle, similar to hemoglobin in blood.

myosin one of two main contractile proteins in a muscle fiber.

myotonic muscular dystrophy the most common adult form of muscular dystrophy marked by muscular rigidity.

neuromuscular junction the point where a nerve fiber and a muscle fiber meet.

neuron the basic cell of the nervous system consisting of a main portion called the cell body, a single long fiber, the axon, and one or more shorter extensions known as dendrites.

neurotransmitter a chemical that spills across the synaptic gap between neurons, transmitting an electrical impulse.

nexus a junction between muscle cells that allows them to pass along electrical signals.

oxygen debt the oxygen consumed in recovery from exercise above the amount that would normally be consumed at rest.

pas de deux a dance for two persons in classical ballet.

pectoral muscles muscles in the chest that form both the pectoralis major, which acts in horizontal flexion, and the pectoralis minor, the prime mover for abduction and downward rotation of the shoulder girdle.

peristalsis the wavelike muscle contractions that push food and other substances through the body's tracts.

peroneus muscle one of several muscles that abduct, evert and flex the sole of the foot.

piloerector a tiny smooth muscle that erects hair follicles.

platysma the platelike facial muscle wrinkling the neck and lowering the jaw.

poliomyelitis a viral disease which, in its acute form, attacks motor neurons in the brain or spinal cord, causing muscle atrophy, deformity or paralysis.

power arm the zone between the fulcrum and the muscle.

primary motor cortex a band of neurons in the frontal lobe of the brain which, when electrically stimulated, induces muscle movements.

prime mover a muscle that is most responsible for movement of a part of the body; an agonist.

proprioceptor one of a variety of sensory receptors located in muscles, tendons, joints and the vestibular organs of the inner ear.

proteinases enzymes that break down protein.

Purkinje fibers part of the network of specialized cells in the heart that spread the impulse to contract muscle fibers in the heart's ventricles.

pyramidal system a major transmission network for movement, consisting of nerve tracts descending from the cerebrum to the spinal cord; important in fine, skilled movements.

pyruvic acid the chemical precursor of lactic acid.

quadriceps the thigh muscle that extends the leg.

reflex an involuntary muscle response such as a hiccup, sneeze or withdrawal of a limb from a painful stimulus.

reflex arc the path traveled by a reflex, where it is converted from an incoming sensory impulse to an outgoing motor impulse.

rhomboid a muscle that rotates the shoulder blade downward.

rigor mortis the muscular stiffness that occurs a few hours after death.

risorius a facial muscle that draws the mouth sideways.

rotator cuff a band of four muscles helping to hold the arm in the shoulder joint.

sarcolemma a thin plasma membrane; the "skin" of a muscle fiber.

sarcomere one of many compartments in a myofibril containing muscle's individual mechanisms of contraction.

sarcoplasm the jellylike substance in a muscle fiber, embedding myofibrils.

sartorius the longest muscle in the body; it assists in flexion, abduction and outward rotation of the thigh at the hip joint and rotates the knee inward.

scalene muscles a group of muscles that act together as the prime mover for side flexion and as assistant mover for flexion of the cervical portion of the spine.

second-class lever system an arrangement of bone and muscle in which the resistance lies between the fulcrum and the muscular force pitted against the burden.

second law of motion the law of physics stating that force is the product of mass and acceleration; one of its principles holds that a moving object will stop or change direction if acted upon by an external force greater than its own.

secondary motor area the cerebrum's cortical region believed to be crucial to intricate movements.

sensory motor integration the constant interaction between sensory feedback and muscle actions that occurs between the spinal cord and the part of the brain that specializes in coordinating sensation and response.

sensory neuron a neuron that conveys impulses to the brain or spinal cord from the body's periphery.

serratus anterior a muscle on the outer surface of the ribs; the prime mover for abduction and upward rotation of the shoulder girdle.

shoulder girdle the skeletal framework attached to and supporting the arms.

sinoatrial node the heart's natural pacemaker containing specialized cardiac muscle fibers that initiate the electrical signals causing the heart to beat.

skeletal muscle the most prevalent type of muscle in the body, usually anchored to bone to carry out voluntary movement.

skeletal muscle fibers cylindrical, multinucleate cells with contractile threads that shorten when stimulated.

sliding filament theory the leading theory of muscle contraction holding that protein filaments in muscle fibers slide past each other when a muscle contracts.

slow-twitch fiber one of the two major types of skeletal muscle fiber; it contracts and fatigues slowly and relies mainly on aerobic energy.

smooth muscle the muscle of the body's internal organs.

somatic sensory region a portion of the cerebral cortex, located in the parietal lobe, where sensory messages are received.

somatotyping a method of classifying human physiques by their degree of fatness, muscularity and leanness, or endomorphy, mesomorphy and ectomorphy.

spinal cord an oval cylinder roughly a half-inch wide within the spinal cavity, containing nerve tracts.

spinal nerves thirty-one pairs of nerves emerging from the spinal cord which relay sensory and motor impulses to and from the skeletal muscles.

sprain a joint injury involving partial or total rupture of ligaments.

steroids a group of naturally occurring and synthetic chemicals that include some hormones, bile acids, hydrocarbons and other substances.

strain damage to muscles or tendons from excessive stretching or use.

striated having stripes.

stroke volume the amount of blood ejected from a ventricle of the heart with one beat.

supraspinatus muscle the small muscle that is a prime mover and first-class lever of the upper arm.

synapse the gap between neurons across which a nerve impulse passes.

synergists muscles that act with other muscles to prevent movements hindering the work of prime movers.

tendon the tough, fibrous tissue that secures each end of a muscle to bone.

tetanic contraction a sustained muscular contraction produced by a rapid succession of impulses.

tetanus a deadly infectious disease that causes paralysis by producing sustained muscular contractions; also a term used to describe normal contractions.

thalamus a large nerve mass crowning the brainstem that conveys motor and sensory impulses to the cerebrum.

thermogram a record, often like a photograph, of the surface temperature of a living body.

third-class lever system an arrangement of bone and muscle in which the force is applied between the fulcrum and resistance.

third law of motion the law of physics stating that to every action there is an equal and opposite reaction.

tibialis anterior the muscle that inverts and flexes the foot back.

tibialis posterior the muscle that inverts and flexes the sole of the foot.

tone the steady, partial contraction of muscles that keeps them ready for action.

transversus abdominis a thick, fibrous muscle crossing the abdomen that constricts abdominal organs and other contents.

treppe the "staircase effect" in which a succession of individual muscle twitches produces greater and greater degrees of contraction.

triceps brachii muscles on the back of the upper arm; a prime mover for extension of the elbow joint.

tropomyosin a long threadlike protein that circles actin filaments and covers binding sites receptive to myosin's cross bridges.

troponin a protein that works along with calcium when a muscle fiber is stimulated, helping to set the contractile mechanism into action.

twitch a simple muscle contraction lasting only a fraction of a second.

visceral smooth muscle one of two types of smooth muscle, made of tightly knit fibers forming most of the body's organs.

Vitruvian man a drawing of the human form according to geometric laws propounded by Roman architect Marcus Vitruvius Pollio; an artistic principle popular during the Renaissance.

volta a popular dance of sixteenth-century England, said to have been a favorite of Queen Elizabeth I.

waltz a lively dance popular in the nineteenth century.

Illustration Credits

It Seems to Dance
6, Michael Abbey/Photo Researchers.

The Body in Motion
8, Rafael Beer © 1980. 10, Mohammed Amin/Bruce Coleman, Inc. 11, Simon Trevor/Bruce Coleman, Inc. 12, (top) Sven Lindblad/Photo Researchers (bottom) detail, *Buffalo Hunt* by William Leigh, National Cowboy Hall of Fame, Photo Researchers. 14, *Discobolo Lancilotti,* SCALA/Editorial Photo Archives. 15, The Granger Collection, New York. 17, (top) *Thumbs Down* by Gerome, The Bettmann Archive (bottom) *Il Puglatore,* SCALA/Editorial Photo Archives. 18, *Dance of the Peasants* by Pieter Brueghel the Younger, SCALA/Editorial Photo Archives. 19, BBC Hulton Picture Library. 20-21, The Bettmann Archive. 22, *Football Players* by Henri Rousseau, The Granger Collection, New York. 24, SNARK/Editorial Photo Archives. 25, Mary Evans Picture Library. 26, *Women in a Coal Mine* by Léonard DeFrance, The Granger Collection, New York. 27, Dr. Nigel Smith/Animals Animals. 28, Smithsonian Institution. 29, *Self Portrait* by Leonardo DaVinci, The Bettmann Archive. 30, *Newton* by William Blake, The Granger Collection, New York. 31, National Library of Medicine. 32, Thomas Eakins, Philadelphia Museum of Art, given by Charles Bregler. 33, **Thomas B. Allen.** 34, (top) The Mansell Collection Ltd. (bottom) *George M. Patchen* by Currier and Ives, Joseph Martin/SCALA/Editorial Photo Archives. 35, Philadelphia Museum of Art, gift of the City of Philadelphia, Trade and Convention Center, Department of Commerce.

The Muscle Machine
36, Historical Collections, College of Physicians of Philadelphia. 38, (top) Lennart Nilsson from his book *Behold Man,* published in the U.S. by Little, Brown & Co., Boston (bottom) Biophoto Assoc. 39, Arthur M. Siegelman/FPG. 40, (top) London Scientific Fotos (bottom) Dr. Victor B. Eichler, Wichita, KS. 41, **Carol Donner.** 42, (top) Arthur M. Siegelman/FPG. 42-43, **Carol Donner.** 44, Dr. Abe Eastwood, College of Physicians and Surgeons, Columbia University. 45, AFIP 82-8572; courtesy of the Armed Forces Institute of Pathology. 46, **Thomas B. Allen.** 47, (top) Dr. Hugh E. Huxley, Medical Research Council, Cambridge, England (bottom) **Jennifer Arnold,** based on the work of Dr. Hugh E. Huxley. 48, (left) Drs. John Heuser and Roger Cooke (right) **Elsie Hennig.** 49, Arthur M. Siegelman/FPG. 50, Tass from Sovfoto. 51, The Bettmann Archive. 52 and 53, Muscular Dystrophy Assn. 54, Annie Leibovitz/Contact Stock Images. 55, Dr. Philip D. Gollnick, Department of Physical Education for Men, Washington State University.

Six Hundred Engines
56, *Murals at the Conference Room: Waterfront Construction* by Thomas Hart Benton, Joseph Martin/SCALA/Editorial Photo Archives. 59, Neil Leifer/*Sports Illustrated.* 60, Jerry Cooke/*Sports Illustrated.* 61, 62 and 63, National Library of Medicine. 64, (top) Ron Modra/*Sports Illustrated* (bottom) **Jennifer Arnold.** 65, (left) Herbert Migdoll (right) Ruth Geyra Stern/Transworld (top and bottom) **Jennifer Arnold.** Foldout, (outside) National Library of Medicine (inside) **Carol Donner** and F. R. Suarez, M. D., consultant, Georgetown University School of Medicine. 66, Children's Hospital Medical Center, Boston, MA. 68, Leonard Kamsler © 1982. 69, Tony Triolo/*Sports Illustrated.* 70, **Thomas B. Allen.** 71, Detail from *Hammerstein's Roof Garden* by William J. Glackens © 1903. Oil on canvas 30x25 inches, collection of the Whitney Museum of American Art.

Circuits for Movement
72, Manfred Kage/Peter Arnold, Inc. 74, National Library of Medicine. 75, Photo courtesy of Dr. C. J. Gibbs, Jr., Deputy Chief, Laboratory of Central Nervous System Studies, NINCDS. 76, Matt Herron/Black Star. 77 and 78, **Mark Seidler.** 79, *The Clubfoot* by José de Ribera, The Granger Collection, New York. 80-81, Fonds Albertina. 82, Lennart Nilsson from his book *Behold Man* published in the U. S. by Little, Brown & Co., Boston. 83, **Christine D. Young.** 84-85, Hirmer Fotoarchiv, Munich. 85, (right) WHO Photo. 86, Micrograph produced by Dr. John E. Heuser of Washington University School of Medicine, St. Louis, MO. 87, Napoleon Chagnon/Anthro Photo. 88, Lawrence Fried/The Image Bank. 89, Paul Wheater/Biophoto Assoc. 90, Ann Ronan Picture Library. 91, **Judith Glick.** 92, **Joyce Hurwitz.** 95, **Thomas B. Allen.** 96, Georg Fischer/Woodfin Camp & Assoc. 97, Michael Inderrieden.

The Language of Movement
98, *Chanteuse au Gant* by Edgar Degas, courtesy of Fogg Art Museum, Harvard University, Bequest-Collection Maurice Wertheim. 100, Nelson Merrifield/FPG. 101, (top) ZEFA. 101, (bottom) and 102, Sybil Sassoon/Robert Harding Picture Library, London. 103, Jack Vartoogian © 1978. 105, Toshio Watanabe/DPI. 106, Royal Academy of Dancing Library, from *The Art of Dancing* by Kellom Thomlinson, 1735. 107, *Murals at the Conference Room: Entertainment and Social Life* by Thomas Hart Benton, Joseph Martin/SCALA. 108, © Herbert Migdoll. 109, Malcolm Hoare/Interlink. 110, Nostovi from Sovfoto. 112, (top) Lois Greenfield/Bruce Coleman, Inc. (bottom) Labanotation by Mary Corey/Dance Notation Bureau. 113, © Dick Zimmerman, Inc. 114, Bill and Claire Leimbach. 115, *Washingtonian.* 116, (top) BBC Hulton Picture Library (bottom) Gallaudet College Photo. 117, *Vitruvian Man* by Leonardo DaVinci, SCALA/Editorial Photo Archives. 118, *Woman Drying Her Neck* by Edgar Degas, SCALA/Editorial Photo Archives. 119, *Nudo per la battaglia di Cascina* by Michelangelo, SCALA/Editorial Photo Archives.

Power and Prowess
120, *Willis* by Joe Wilder, courtesy of Spectrum Fine Art, Ltd., New York. 122, Seth Eastman: *Lacrosse Playing Among the Sioux Indians,* in the collection of The Corcoran Gallery of Art, gift of William Wilson Corcoran. 123, Focus on Sports. 124, Howard Sochurek © 1982. 126, **Thomas B. Allen.** 127, Ted Streshinsky © 1980. 128, Gerald Brimacombe/The Image Bank. 129, Heinz Kluetmeier/*Sports Illustrated.* 131, Drs. R. E. Burke and P. Tsairis, National Institutes of Health, Bethesda, MD. 132, Courtesy of Dr. Carl Klafs. 133, National Library of Medicine. 134, Lane Stewart/*Sports Illustrated.* 135, Christopher Morrow. 136, Ralph Morse from *Scientific American.* 138, Bernard Pierre Wolff/Magnum. 139, Douglas Kirkland/Contact Stock Images. 141, Ted Streshinsky © 1980. 142-143, Tim Morse/Morse Photography. 144, *Business Men's Bath* by George Bellows, courtesy of W. H. Allison & Co., Inc. 145, Wasyl Szkodzinsky/Photo Researchers. 146, From the Photo Archives of the U. S. Olympic Committee. 147, W. Iooss/The Image Bank. 148, New York Times Pictures. 149, Mel Digiacomo/Focus on Sports. 150, World Wide Photos, Inc. 151, From the Photo Archives of the U. S. Olympic Committee. 152, Joseph Daniels. 153, Ian Berry/Magnum.

Appendix
154 and 155, **Donald Gates.**

Index

Page numbers in bold type indicate location of illustrations.